C000142197

Edexcel AS | 1

UNIT

4/2010

Geography

Global Challenges

Cameron Dunn and Sue Warn

Philip Allan Updates, an imprint of Hodder Education, an Hachette UK Company, Market Place, Deddington, Oxfordshire OX15 0SE

Orders

Bookpoint Ltd, 130 Milton Park, Abingdon, Oxfordshire, OX14 4SB
tel: 01235 827720
fax: 01235 400454
e-mail: uk.orders@bookpoint.co.uk
Lines are open 9.00 a.m.–5.00 p.m., Monday to Saturday, with a 24-hour message answering service. You can also order through the Philip Allan Updates website: www.philipallan.co.uk

© Philip Allan Updates 2008

ISBN 978-0-340-94931-3

Impression number   5
Year   2013   2012   2011   2010   2009

This Guide has been written specifically to support students preparing for the Edexcel AS Geography Unit 1 examination. The content has been neither approved nor endorsed by Edexcel and remains the sole responsibility of the author.

Printed by MPG Books, Bodmin

Hachette UK's policy is to use papers that are natural, renewable and recyclable products and made from wood grown in sustainable forests. The logging and manufacturing processes are expected to conform to the environmental regulations of the country of origin.

Contents

Introduction

■ ■ ■

Content Guidance

■ ■ ■

Questions & Answers

Introduction

About this guide

This guide is for students following the Edexcel AS Geography course. It aims to guide you through **Unit 1: Global challenges**.

This **Introduction** provides an overview of the unit, introduces the specification content and summarises the assessment.

The **Content Guidance** section provides a detailed guide to 'World at risk' and 'Going global'.

The **Questions & Answers** section provides examples of two Section A questions, with outline mark schemes and exam tips on how to improve your performance in the exam. It also provides examples of two Section B questions, again with outline mark schemes and C-grade student responses with commentary to show how they could be improved.

Overview of Unit 1

Unit 1 examines issues associated with 'World at risk' and 'Going global'. This guide provides you with a summary of the key concepts and structures you need to study the unit effectively and to revise for the exam.

The headings for each of the 14 sections follow the specification closely. In the specification, the left-hand column is what you need to learn — this is the one the examiners will use when setting questions in the exam. The right-hand column suggests how you might study the concepts and gives you a context for your studies.

As well as using your textbook, you need to keep up to date by following key themes such as Polish migration, global trends in hazards or the latest information on climate change, using articles in newspapers and websites.

While the unit looks at issues on a global scale, there are several opportunities for you to carry out locally based research, such as exploring the roots of your community, assessing local hazard risks or looking at how you and your family could reduce your carbon footprint.

This unit has six compulsory in-depth case studies. These are:
- Global warming in the Arctic (1) and continent of Africa (2)
- Multiple hazard hotspots in the Philippines (3) and California (4)
- Migration flows — eastern Europeans to the UK (5), and UK retirement migration to the Mediterranean (Spain) (6)

Apart from these case studies, which you need to learn in detail, you will be better off learning shorter examples to use in the exam, so that you can support and develop your arguments. It is also well worth building up a vocabulary list of all the key words you learn over the course to improve your understanding of the concepts. The glossary or key terms in your textbook will give you some good ideas.

Assessment

In the examination Section A tests the *breadth* of your knowledge and understanding across the whole unit. It is marked out of 65 marks and you have to complete six short structured questions in under 1 hour. A typical question includes, in increasing order of difficulty:

- definition of a term (1 mark)
- multiple choice (1 mark)
- simple analysis of a resource (2 marks)
- development of a resource (3 marks)
- an explanation based on development from a resource (4 marks)

Questions in Section A are marked in points.

Section B tests the *depth* of your knowledge and understanding. You have to choose one question from the four set, worth 25 marks, and you have around 30 minutes to answer it. Each question consists of two parts:

- data stimulus (10 marks)
- extended writing, using your compulsory case studies (15 marks)

Section B questions are marked using levels, and the quality of written communication is assessed.

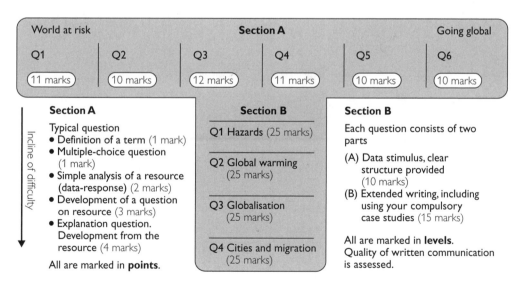

World at risk		Section A			Going global
Q1	Q2	Q3	Q4	Q5	Q6
11 marks	10 marks	12 marks	11 marks	10 marks	10 marks

Incline of difficulty

Section A

Typical question
- Definition of a term (1 mark)
- Multiple-choice question (1 mark)
- Simple analysis of a resource (data-response) (2 marks)
- Development of a question on resource (3 marks)
- Explanation question. Development from the resource (4 marks)

All are marked in **points**.

Section B

Q1 Hazards (25 marks)

Q2 Global warming (25 marks)

Q3 Globalisation (25 marks)

Q4 Cities and migration (25 marks)

Section B

Each question consists of two parts

(A) Data stimulus, clear structure provided (10 marks)
(B) Extended writing, including using your compulsory case studies (15 marks)

All are marked in **levels**. Quality of written communication is assessed.

Content
Guidance

Unit 1: **Global challenges** consists of two compulsory topics:
- **World at risk**
- **Going global**

The unit focuses on the issues of global hazards and globalisation at the broad scale. However, it also provides smaller-scale case studies and opportunities to explore how your life interacts with these global challenges.

This Content Guidance section summarises the key information you need to know for each of these topics. The information for each topic is divided into subsections that match those of the specification.

World at risk:
- Global hazards
- Global hazard trends
- Global hazard patterns
- Climate change and its causes
- The impacts of global warming
- Coping with climate change
- The challenge of global hazards for the future

Going global:
- Globalisation
- Global groupings
- Global networks
- Roots
- On the move
- World cities
- Global challenges for the future

World at risk

'World at risk' contains two interlinked strands — **global hazards** and **climate change** — which together increase the **risk** from disasters, especially for the world's most **vulnerable** people.

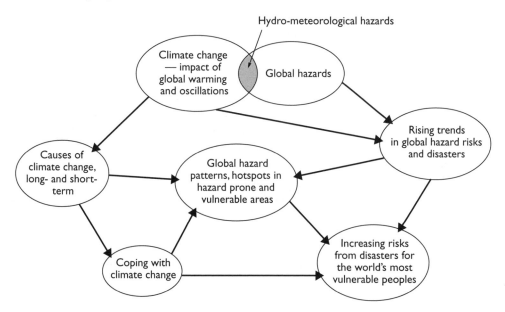

Figure 1 Topic concept map

Global hazards, which can be classified into two major categories, **hydro-meteoro-logical** and **geophysical**, cause increasing numbers of disasters around the world. Frequently a number of hazards occur together (i.e. are **spatially concentrated**). Where the people are already vulnerable, because of poverty and high population densities, **multiple-hazard hotspots** occur. A study of **global trends** confirms that the **frequency** and **magnitude** of hydro-meteorological disasters is increasing as a result of a possible link to **climate change**.

Climate change is considered by many to be the world's greatest problem. Technically, it is a **context hazard** and therefore a 'chronic global-scale threat' to both the environment and people.

This topic explores the causes and impacts of climate change and solutions to the problem. It considers whether recent global warming is unprecedented in the history of climate change and is therefore **anthropogenically** caused. Short-term climate change encompasses not only global warming but also the impact of **oscillations** such as El Niño.

Global hazards

Risk and vulnerability in an unequal world

- **Environmental hazards** can be defined as 'the threat posed to humans and the environment by natural events originating in it'. These natural events can be classified in many ways, but it is useful to categorise them as geophysical or hydro-meteorological. Humans invariably make the event worse and there are very few hazards that are caused by purely physical factors.
- **Geophysical hazards** are formed by tectonic/geological events (earthquakes, volcanoes and tsunamis).
- **Hydro-meteorological hazards** are formed by hydrological (floods) and atmospheric (storms and droughts) processes.
- **Context hazards** are widespread threats arising from global environmental changes such as climate change or from a major hazard such as a super-volcano.
- **Disasters** are the 'realisation' of hazards to cause social impacts such as loss of lives and livelihoods, and economic impacts such as damage to goods and property (i.e. the actual *consequences* of a hazard).
- **Risk** is the probability of a hazard occurring and becoming a disaster as a result of deaths and loss of livelihoods, goods and property (i.e. the likely *exposure* to a hazardous event).
- **Vulnerability** is shaped by the underlying state of human development, including inequalities of income, opportunity and political power, all of which marginalise the poor. Poverty-stricken people frequently depend directly on the land for food (subsistence), live in fragile eco-environments and have poor health and well-being. The result is that high risk is combined with an inability to cope with the adverse impact of natural hazards and climate change (at all levels — individuals, communities and countries). These relationships are summarised in Figure 2.

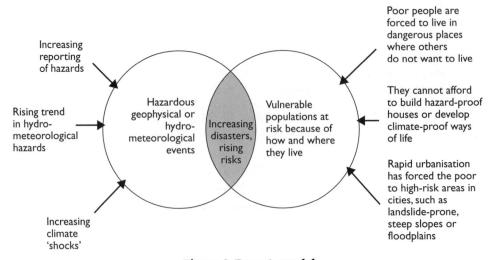

Figure 2 Dregg's model

Risks and vulnerability are now greater as a result of climate change, with more hydro-meteorological disaster events, which add to the pressure on countries that have a limited ability to adapt and where coping strategies are already stretched.

Rising risks

The **risk equation** measures the level of hazard for an area:

$$\text{risk } (R) = \frac{\text{frequency or magnitude of hazard } (H) \times \text{level of vulnerability } (V)}{\text{capacity of population to cope and adapt } (C)}$$

$$R = \frac{H \times V}{C}$$

The risk is getting worse for many communities and countries because:

- the frequency and magnitude of hazards are increasing with climate change
- vulnerability is increasing as a result of unsustainable development leading to poor land use and environmental degradation
- the capacity to cope is decreasing owing to poverty and urbanisation

Note that the risk is much lower in economically developed countries. These countries have the resources and technology to provide protection from the worst impacts of disasters.

Recent climate change: the world's greatest hazard?

It is important to use the correct terminology — see the 'Climate change survival guide' below. Climate change can be classified using timescales (long, medium and short term). Short-term or recent climate change, most marked since the 1960s, is not only the result of global warming but also of the increasing impact of atmospheric/oceanic oscillations such as the El Niño/La Niña effect.

> **Exam tip**
>
> Be careful not to blame every extreme weather event on global warming. Atmospheric/oceanic oscillations are a more likely culprit. Global warming tends to affect short-term climate trends.

Climate change survival guide

- **Climate change** is any marked trend or shift in climate (average weather over 30 years) that shows a sustained change in the average value for any particular climatic element (rainfall, drought, storminess etc.) Clearly, for longer- and medium-term climate change, evidence other than records of weather measurement has to be used.
- **Climate variability** refers to the differences in climate from one year to another. More extreme weather seems to be associated with the current climate.
- **Global warming** refers to a consistent recently measured rise in the average surface temperature of the planet (i.e. at the global scale). However, note that neither the rise in temperature nor the related impacts on atmospheric and oceanic circulation are likely to occur uniformly across the globe.

- The **greenhouse effect** is a natural process that warms the Earth's atmosphere due to the trapping of heat that would otherwise be radiated back into space — it enables the survival of life on Earth.
- An **enhanced greenhouse effect** occurs when the amount of greenhouse gases (the main ones are water vapour, carbon dioxide, methane, nitrogen oxide and CFCs) increases. Scientists argue that this is almost certainly the result of anthropogenic (human) activity such as the burning of fossil fuels (oil, coal, gas) and it leads to global warming.
- **Fossil fuels** are rich in carbon and burning them releases carbon dioxide.
- **The tipping point** is the point at which a system switches from one state to another. In terms of climate change, scientists have identified this tipping point as a 2°C rise in temperature. This small rise in temperature could be enough to lead to dramatic and possibly cataclysmic changes in the environment which are probably irreversible. For example, the failure of the Gulf Stream to reach northwest Europe would contribute to a much more hazardous world.
- **The 'hockey stick' graph** shows the average global temperature change going back 1,000 years. It is so called because it shows a long flat period followed by a rapid increase in temperature in the last 50 years.
- **A feedback mechanism** is a process whereby the output of a system acts to amplify (positive) or reduce (negative) further output. For example, the melting of Arctic permafrost would lead to the release of large quantities of trapped methane, which would itself lead to further global warming.
- **Global dimming** is the phenomenon whereby particles in the atmosphere, largely from industrial pollution, have led to less solar radiation reaching the Earth (in fact this may have masked the initial 1960s global warming).
- **IPCC** is the **Intergovernmental Panel on Climate Change**, which was set up in 1988. It is the central scientific voice on the issue.

Ten reasons why recent climate change might be the world's greatest hazard

1 Climate change is a **global problem**: all areas of the world will be affected to a lesser or greater degree. However, the impacts are various, and often unique to a particular country.
2 Climate change is a **chronic (ongoing) hazard**. It will have an enormous range of direct impacts linked to rising temperatures, which can themselves amplify other problems. These include climate belt migration towards the poles, which will have a severe impact on ecology and wildlife and lead to the spread of diseases such as malaria. Rising temperatures also have an impact on both atmospheric and oceanic circulations. Rising ocean temperatures may be a cause of more frequent atmospheric/oceanic oscillations (such as El Niño) and increasing frequency and magnitude of hurricanes.
3 Climate change is not just an environmental problem. It has a **range of impacts** on societies and economies, threatening their very existence in terms of water availability, food security and health and wellbeing.

4 Climate change is **unpredictable**. It could become catastrophic depending on whether, or how fast, the tipping point is reached. This depends on both physical and human factors:

- There are a number of interlocking systems that can affect each other via **positive** and **negative feedback**. For example, rising temperatures melt glaciers and ice sheets. This leads to more incoming solar radiation being absorbed by land and sea, because ice and snow have a greater albedo (reflective power) than rock and water. Thus melting of the ice sheet leads to yet more melting — an example of positive feedback. The melting ice sheets could lead to the ocean becoming diluted, which could have an effect on ocean current circulation, thus weakening the warming powers of the North Atlantic Drift (negative feedback).
- The human factor is related to how effectively the world can **mitigate** the damaging effects of climate change by reducing the production of greenhouse gases to slow down the rate of global warming. Many people argue that only with radical solutions will the impacts of climate change be halted.

5 Climate change has many **indirect impacts** too. For example, thermal expansion of the oceans as water expands with heating leads to rising sea levels, a process that may reach catastrophic levels if the large polar ice sheets (Arctic and Antarctica) melt. Rising sea levels have already created the world's first 'environmental refugees' in the islands of the Ganges delta and the South Pacific.

6 Climate change prediction requires successful **modelling** of the future by scientists in the IPCC for both direct impacts (rising temperatures, more extreme weather) and indirect impacts (rising sea levels, changing ecosystems). The sheer complexity of the calculations makes this process extremely difficult, with huge variations in the scenarios presented. This makes it particularly hard to convince people about the severity of climate change.

7 It is difficult for scientists and researchers to **separate the effects** of global warming from those of other influences on present-day weather such as the El Niño/La Niña cycle. Many of the changing weather patterns — such as more extremes with greater climate variability, or increasing frequency of droughts, big storms and widespread flooding — may be attributable to global warming, but are more likely to be linked to atmospheric/oceanic oscillations. The droughts and floods in 2007 are an example.

8 Climate change requires **global solutions**, primarily in tackling carbon emissions. However, there are huge political problems. It is unjust to have a two-speed, unequal world in which more developed countries pollute and less developed countries are vulnerable victims. A further problem is the emergence of China, India and Brazil as rapidly expanding economies, and a resurgent Russia, which feel that their industrial development should not be jeopardised by targets for reducing greenhouse gas emissions.

9 Until recently the issue of climate change was **strongly contested**. While there is now a broad consensus among scientists, led by the IPCC, that levels of greenhouse gases and global temperatures are rising and that there is a link between them, there are still sceptics. There are two such groups: those who find weaknesses in the data and their interpretation; and those with vested interests (for example, oil companies), which claim that global warming and the resulting hazards are due to factors other than the burning of fossil fuels. However, the awarding of the 2007 Nobel peace prize to Al Gore (who presented the film on global warming *An Inconvenient Truth*) and to the Chair of the IPCC gave legitimacy to the campaign to take action on climate change.

10 Climate change is a problem of enormous scale and proportions, and so is the world's most **costly problem**.

Global hazard trends

Is the world becoming more hazardous?

The issue

To find out whether the world is becoming more hazardous or not, you need to use statistics from key websites (see below) to discover whether more hazard events are occurring and, if so, of what type and why. Remember that many textbooks use the terms 'hazard' and 'disaster' interchangeably. There may not be more hazards of all types occurring, but a higher percentage are becoming disasters because they affect increasing numbers of vulnerable people. There is also the fact that hazards and disasters are hot news, and scientists, engineers, governments, insurance companies and risk managers all take great interest in them. It may be that there are not actually more hazards, but more are being reported because of this media interest and the improvements in information and communications technologies.

> **Exam tip**
>
> It is up to you to use the various databases and to explore the evidence.

Disaster statistics

Disaster statistics come from a number of sources. Governments report disaster statistics to UN agencies, e.g. WHO (**www.unisdr.org/disaster-statistics**). These quasi-official statistics are supplemented and cross-checked by designated groups that monitor media and internet reports and also field data from NGOs working on the front line.

The pre-eminent authority is CRED (Centre for Research on the Epidemiology of Disasters, **www.cred.be**), supported by the World Bank. It administers the key database: **www.emdat.net**. Both Munich Re (**www.munichre.com**) and Swiss Re (**www.swissre.com**), which are re-insurance multinationals, provide disaster databases for their own internal use.

> **Exam tip**
>
> Research the databases to get the latest years statistics, as credit is given for up-to-date research.

Trends in the number of reported disasters

- The number of reported natural disasters has risen sharply since the 1960s.

- This trend is particularly marked for **hydro-meteorological disasters**. The increase in floods and storms, and more recently in droughts, is responsible for driving the overall trends. Floods and storms together accounted for 70% of all disasters globally in the last 40 years.
- **Geophysical disasters** such as earthquakes and volcanoes have increased slightly since the 1970s, possibly as a result of more reporting of disasters and also because of the rising number of vulnerable populations living in hazard-prone areas. However, in general, the trend fluctuates. Indonesia is currently a 'hotspot' for tectonic activity, and it has two very active volcanoes. It has experienced five major earthquakes and several tsunamis since 2005. All of these events are the result of a highly mobile plate boundary where the Indian plate is being subducted beneath the Burma plate (at a destructive boundary).
- **Biological hazards** (bio-hazards) have also shown an upward trend from a low base, often numerically exceeding geophysical hazards. This may be linked to global warming which leads to more pests and diseases.

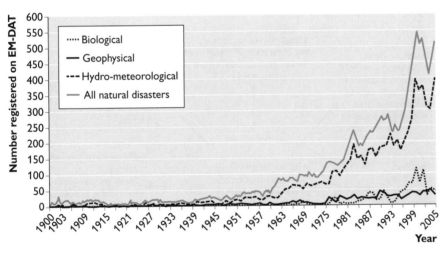

Figure 3 Hazard trends, 1900–2005

The reasons for the rising trend in hydro-meteorological disasters are complex and include both physical and human factors.

The media almost always state that the rising trend is a result of global warming, suggesting for example that rising temperatures lead to warmer oceans, thus 'spawning' more hurricanes, and cause stronger convectional cells — again creating more hurricanes. Equally, rising temperatures increase evaporation, which in turn contributes increasing rainfall and therefore more flooding.

The IPCC suggests that a more extreme climate (resulting from global warming) will increase the risk from certain hazards. In some cases, it may lead to more severe events (greater magnitude); in other cases, the events may occur more often (higher frequency).

Of equal importance are oscillations such as the El Niño/La Niña cycle. The cycle, which affects winds and ocean currents, lasts for 7–10 years, in which 1–2 years are El Niño years and 1–2 years are La Niña years. The cycle has a variety of effects leading to hazardous weather in many parts of the world (these links are known as tele-connections):

- During El Niño years there is unusually heavy rain in Peru and California, and dry conditions in Indonesia and Australia (2006 drought).
- During La Niña years there is unusually heavy rain in Australia and Indonesia, and drought in the Americas (Californian drought and wildfires) and various parts of Africa.
- Hurricanes in the Atlantic are usually increased in La Niña years and reduced in El Niño years.

Some scientists even suggest that global warming may increase the frequency and intensity of these oscillations.

Purely human factors also play a significant part in rising trends in disasters. Figure 4 summarises these factors.

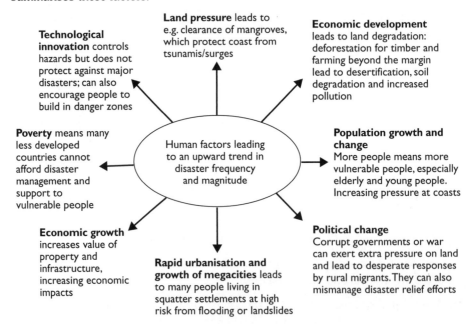

Figure 4 Human factors increasing disasters

Trends in the social and economic impacts of disasters

The statement 'Deaths are falling, whereas the number of people affected by disasters is rising and economic losses are escalating' summarises the situation globally. However, economic development also plays a part. A general statement such as 'Less developed countries experience more deaths (social costs) and more developed countries experience more damage (economic costs)' has some truth.

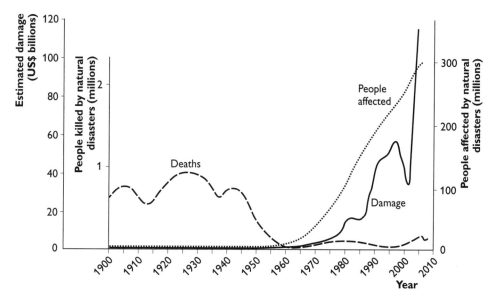

Figure 5 Global trends in deaths, people affected and economic losses from natural disasters, 1900–2006

Falling deaths?

As Figure 5 shows, the number of people reported killed by natural disasters dropped dramatically before the 1970s, but it has levelled off in recent years to a fluctuating trend, with a rise in 2004–05 (mega events). This is largely the result of better disaster management before, during and immediately after the disaster:

- **Before a disaster** there is better mitigation (use of hazard-proof/climate-proof structures) and improved disaster preparedness using education, community training and high-technology warning systems.
- **During the disaster** — in many cases emergency responses have improved, with local participation in effective disaster relief programmes (e.g. Hurricane Sidr in Bangladesh in 2007).
- **Post-disaster** recovery programmes are also better coordinated, often using GIS to pinpoint problems. Studies of the Kashmir earthquake (2005), Hurricane Katrina (2005), Typhoon Nargis (2008) and the south Asian tsunami (2004) do however highlight many problems, but there is no doubt that, especially in developed areas, deaths are minimised by an effective disaster response. In the Sichuan earthquake in 2008 the Chinese government attempted a prompt disaster response in spite of numerous difficulties.

All this is against a backdrop of increasing hazards and disasters, especially from hydro-meteorological causes.

Rising numbers of people affected

The definition of 'being affected' includes loss of home, crops, animals, livelihood, and often a decline in health/quality of life for a designated period (usually 3 months).

As Figure 5 shows, in the 1970s on average 100 million people a year were affected by disasters, but in the early twenty-first century this has risen to over 300 million a year (with peaks up to nearly 700 million as a result of the 2004 tsunami and other mega disasters — not shown on the graph). Increased vulnerability due to a less favourable risk equation (see page 11) is responsible for this rising trend, as some 80% of the millions affected are concentrated in areas of low and low-to-medium human development. Countries with the greatest numbers of people affected are almost exclusively less developed countries and new industrialising countries (NICs) such as India and China.

Economic losses from disasters

Economic losses from disasters are growing at a faster rate than the number of disasters. The value of the losses is escalating compared with the insured risks, because hazard-prone areas in general have high insurance premiums, and some businesses and many poor people (who are forced to live in hazard-prone areas) cannot afford or even access insurance. Rising numbers of disasters are occurring in less developed countries where insurance is rarely an option. Mega-disasters such as Hurricanes Andrew and Katrina can lead to huge rises in economic losses for a single year, resulting in a fluctuating but nevertheless overall rising trend.

There is a tendency to over-emphasise the economic damage and losses experienced in more developed countries. In absolute terms, there is no doubt that because of the value of possessions, the sophistication of installations and infrastructure, and the scale of domestic insurance claims, the amounts involved in developed countries are large.

However, in relative terms, economic damage from natural disasters tends to be higher in less developed countries, mainly because of their high dependence on one or two economic sectors, such as tourism or cash cropping of bananas. The damage forms a very high percentage of their very low annual gross domestic product (GDP). For example, Hurricane Mitch in 1998 caused a 50% loss in GDP for Nicaragua. By contrast, Munich Re estimated that between 1994 and 2003 Japan suffered $166 billion of economic damage from national disasters but this was only 2.6% of its GDP.

Another reason why the rate of economic losses is increasing faster than the occurrence of disasters is the growing economies of NICs such as India, China and the 'Tiger' economies of Korea, Taiwan, Singapore and Malaysia.

Global hazard patterns

Exploring local hazards and links to climate change

You must research hazard risks in your local area or another small-scale area. A starting point is your local reference library. Research local records including newspaper archives and websites that record historic and present-day hazard events. You will probably find rising numbers of floods and storms, but less snowfall.

When you are researching climate change you will find it is difficult to get **climatological data** on changing climates for your local area, but there are a number of websites that provide this information at a regional or national scale. The Climate Research Unit (**www.cru.uea.ac.uk**) and the Tyndall Centre (**www.tyndall.ac.uk**) provide useful maps and data for the UK from which you can extract information for your local area.

Another interesting method of research is the science of **phenology,** which looks at key indicators to measure the response to climate change. Figure 6 summarises some of these indicators. All the results seem to indicate that in recent years (especially 2007) we have had warmer winters and earlier springs. Natural events such as those shown may have moved up to 15 days earlier over the last 30 years. There are a number of useful websites for this type of research (e.g. **www.naturescalendar.org.uk**, **www.cru.uea.ac.uk/cru/info/iccuk** and **www.ukcip.org.uk**).

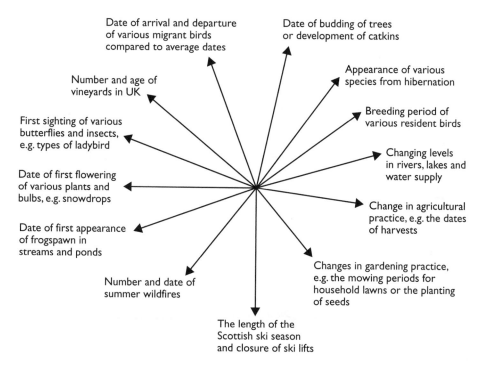

Figure 6 Possible climate change response indicators

The distribution of the world's major natural hazards
For each of the major hazards (earthquakes, volcanoes, slides, drought, floods and storms), you need to know the following:

- The **distribution** of the occurrence for each hazard. A world map to show this is crucial, as the distributions for each hazard are very different.
- A clear **definition** of each of the six types.
- A basic understanding of the **causes** of each hazard. For example, for geophysical hazards you need to know how plate tectonics affects the distribution of earthquakes and volcanoes.

- Some understanding of the likely **impacts** of each hazard, so that you can weigh up its contribution to disaster statistics (such as deaths and damage) and also see how it affects your case studies of disaster hotspots.

The distribution of earthquakes

As can be seen from Figure 7 the main earthquake zones are clustered along plate boundaries. The most powerful earthquakes are associated with destructive or conservative boundaries. Earthquakes are a common hazard and can develop into major disasters when they are of high magnitude (i.e. Richter scale 6+) and occur in a densely populated/vulnerable area (e.g. Kashmir, 2005). Earthquakes not only cause primary hazards from ground movement and ground shaking, but also cause secondary hazards such as landslides and even tsunamis (e.g. south Asian tsunami, 2004).

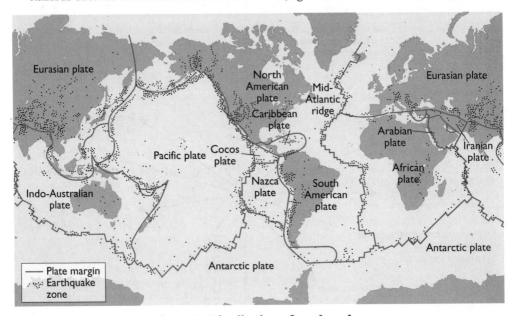

Figure 7 Distribution of earthquakes

Destructive plate boundaries
Destructive boundaries, where oceanic crust is being subducted beneath a continental plate (e.g. Philippines case study), or where two oceanic or continental plates collide, produce shallow, intermediate and deep earthquakes.

Constructive plate boundaries
Constructive boundaries (where oceanic plates are moving apart) are associated with large numbers of shallow, low-magnitude earthquakes. Most are submarine (under the sea).

Conservative plate boundaries
Conservative boundaries, where the plates slide past each other, produce frequent, shallow earthquakes, sometimes of high magnitude (e.g. along the San Andreas fault system of the western USA).

Other earthquakes
A small minority of earthquakes occur within plates, usually involving the reactiva-tion of old fault lines (e.g. the Church Stretton fault in Shropshire). Occasionally earth-quakes can result from human actions, such as dam or reservoir building, which increase the weight and therefore stress on the land (e.g. in Killari, northern India, in 1993, an earthquake caused by a dam killed 10,000 people).

The distribution of volcanoes
Figure 8 shows the global distribution of active volcanoes. The type of tectonic situa-tion determines the composition of the magma and therefore the degree of explo-sivity of the eruption, which is a key factor in the degree of hazard risk.

Constructive plate boundaries.
Most of the magma that reaches the Earth's surface wells up as volcanoes at ocean ridges, such as the mid-Atlantic ridge. These volcanoes are mostly on the sea floor and do not represent a major hazard to people except where they emerge above sea level to form islands (e.g. Iceland). The east African rift valley, which is a construc-tive plate boundary, has a line of 14 active volcanoes, some of which can produce dangerous eruptions (e.g. Mt Nyragongo in the Democratic Republic of Congo, 2002).

Destructive plate boundaries
Some 80% of the world's most active volcanoes occur along destructive boundaries. When oceanic plates are subducted beneath continental plates, explosive volcanoes, such as Mt St Helens, are formed. The 'ring of fire' around the Pacific has many such volcanoes, including those in the Philippines.

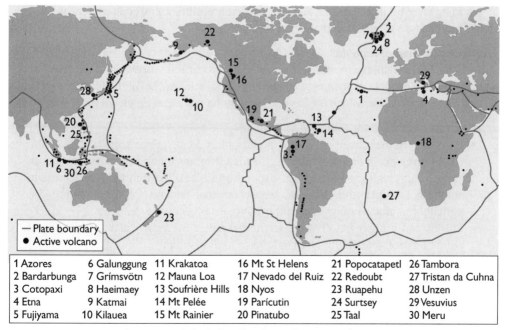

1 Azores	6 Galunggung	11 Krakatoa	16 Mt St Helens	21 Popocatapetl	26 Tambora
2 Bardarbunga	7 Grímsvötn	12 Mauna Loa	17 Nevado del Ruiz	22 Redoubt	27 Tristan da Cuhna
3 Cotopaxi	8 Haeimaey	13 Soufrière Hills	18 Nyos	23 Ruapehu	28 Unzen
4 Etna	9 Katmai	14 Mt Pelée	19 Parícutin	24 Surtsey	29 Vesuvius
5 Fujiyama	10 Kilauea	15 Mt Rainier	20 Pinatubo	25 Taal	30 Meru

Figure 8 Global distribution of active volcanoes

Hotspots

Hotspots are localised areas of the Earth's crust (lithosphere) which have an unusually high heat flow, and where magma rises to the surface as a plume (e.g. Hawaii). As a lithospheric plate moves over the hotspot, a chain of volcanoes is created.

Volcanic hazards

Apart from the local impacts of lava flows the most catastrophic impacts of volcanoes are pyroclastic flows, ash falls, tsunamis and mudflows (e.g. in 1983 mudflows from Nevado del Ruiz killed over 22,000 people).

The distribution of slides

Slides include a variety of mass movements, such as rock slides, debris flows, snow avalanches, and rainfall- and earthquake-induced slides.

Landslides

Landslides are the seventh biggest killer with over 1,400 deaths per year, ranking above both volcanoes and drought. Most landslides occur in mountainous areas, often after abnormally heavy rain and/or seismic activity. Human factors also play a part. Deforestation of hillsides in southeast Asia and building on hillslopes in Hong Kong have both led to widespread slides following rain.

Snow avalanches

Snow avalanches are concentrated in high mountainous areas such as the Southern Alps of New Zealand or the Rockies of North America. Avalanches tend to occur on slopes steeper than 35°. An average of 40 deaths a year in Europe and over 100 in North America are caused by avalanches. Recent research has suggested that global warming may be increasing avalanche occurrence, although trends in deaths have slowed because of effective management.

The distribution of drought

Drought has a dispersed pattern — over one-third of the world's land surface has some level of drought exposure (see Figure 9). This includes 70% of the world's people and agricultural value, which means that drought has an effect on global food security.

The causes of drought include the following:

- Variations in the movement of the inter-tropical convergence zone (ITCZ). As the ITCZ moves north and south through Africa it brings a band of seasonal rain. In some years, high-pressure zones expand and block the rain-bearing winds. In southern Ethiopia and Somalia, where farmers depend for food on rain-fed agriculture, famines may result if the summer rains never arrive.
- El Niño can bring major changes to rainfall patterns. In particular (as in 2006), it can bring drought conditions to Indonesia and Australia.
- Changes in mid-latitude depression tracks. In temperate regions, depressions bring large amounts of rainfall. However, if blocking anticyclones form and persist, depressions are forced to track further north, leading to very dry conditions. Droughts in the UK and France (1976, 1989–92, 1995, 2003 and 2006) as well as in the US midwest in the 1930s were all related to this cause.

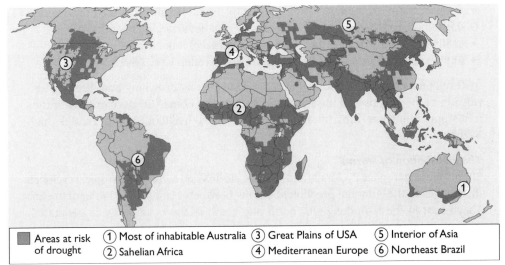

| Areas at risk of drought | ① Most of inhabitable Australia | ③ Great Plains of USA | ⑤ Interior of Asia |
| | ② Sahelian Africa | ④ Mediterranean Europe | ⑥ Northeast Brazil |

Figure 9 Global distribution of drought-risk areas

The distribution of flooding

Flooding is a frequent hazard and is evident in some 33% of the world's area, which is inhabited by over 80% of its population (Figure 10). Regional-scale, high-magnitude floods are frequent events in India/Bangladesh and China.

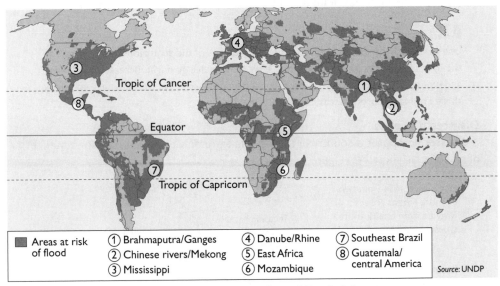

Areas at risk of flood	① Brahmaputra/Ganges	④ Danube/Rhine	⑦ Southeast Brazil
	② Chinese rivers/Mekong	⑤ East Africa	⑧ Guatemala/ central America
	③ Mississippi	⑥ Mozambique	

Source: UNDP

Figure 10 Global distribution of flood-risk areas

The causes of flooding include the following:

- By far the most common cause is excessive rainfall related to atmospheric processes, including monsoon rainfall and cyclones. In temperate climates, a series of depressions sometimes brings prolonged heavy rainfall.

- Intense rainfall sometimes associated with thunderstorms can lead to localised flash flooding. These sudden floods can have a devastating effect.
- El Niño can bring devastating floods, as in Mozambique in 1997 and 2006.
- Rapid snowmelt can add water to an already swollen river system.

In developing countries flooding may lead to deaths by drowning and disease, destruction of food crops and infrastructure and loss of homes. In developed countries it disrupts transport and infrastructure, damages livelihoods and creates high insurance costs.

The distribution of storms

Storms include tropical cyclones, mid-latitude storms and tornadoes. Tropical cyclones (hurricanes in the Atlantic) are violent storms between 200 and 700 km in diameter. They occur in the latitudes 5–20° north and south of the equator. Once generated, cyclones tend to move westward.

Tropical cyclones or hurricanes will only occur over warm ocean (over 26°C) of at least 70 m depth at least 5°N or 5°S of the equator in order that the Coriolis effect (very weak at the equator) can bring about rotation of air.

Storms cause damage in several ways, including heavy rain (leading to floods and mudslides), high wind velocity and very low central pressure (leading to storm surges and coastal flooding). They can be devastating (e.g. Hurricane Katrina and Cyclone Nargis in low-lying mega-delta regions).

Exam tip

Use case-study cards to summarise the distributions of the six major hazard types. For the hydro-meteorological hazards, note how climate change might increase their impact and where. Try to build up your own world map using overlays to see where you think there are hazard-prone areas and likely hotspots.

Disaster hotspots

A **disaster hotspot** is a country or area that is extremely disaster-prone for a number of reasons, as shown on Figure 11.

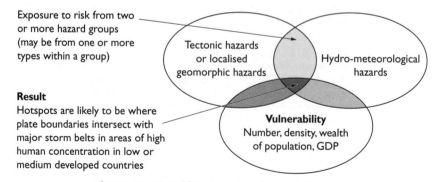

Figure 11 Identification of a disaster hotspot

You need to learn the Philippines and California coast case studies in depth with detailed statistics. Learn how to draw a simple sketch map on which you can annotate key facts about the occurrence and impacts of the range of disasters they experience. For each one, think about the types of hazard and why they are concentrated there (with examples of recent events and the deaths and damage caused). Consider how vulnerability can lead to disasters, and the adaptive capacity of the people and government. The interaction between people and the natural events is the key to understanding their impact.

These two case-study areas have been selected because of the contrast in the types of hazard they experience and their states of development.

Table 1 summarises the contrasts between the two areas.

Table 1 The Philippines and California coast compared

	Philippines	California coast
Status	Lower-middle income country	Wealthiest state of the USA
Population	91 million (density 240 per km^2, 2,000 per km^2 in Manila)	40 million (2008)
GDP per capita	$5,000	$40,000
HDI	0.75	0.95 (nearly the highest)
Landscape	Mountainous country, with crowded coastal lowlands. Consists of 7,000 islands, many very small, spread over latitudes between 5°N and 20°N (typhoon belt)	Coastal area contains the huge conurbations of Los Angeles, San Francisco and San Diego
Summary	Classic hotspot where typhoon belt interacts with mobile plate boundary in a rapidly developing country; El Niño/La Niña cycles increase range of hazards	Elongated area ranging from subtropical in north to tropical in south; major tectonic hotspot, which is also subject to extreme weather hazards brought by El Niño/La Niña oscillations

Compulsory case study: the Philippines
- Typhoons are the main hazard (20 per year).
- Volcanic eruptions are explosive with dangerous lahars (mudflows).
- Earthquakes are common (100% of country at risk).
- Occasional droughts associated with El Niño years.
- Landslides are common in mountain areas.
- The Philippines is a densely populated, rapidly developing country.
- Vulnerability is increased by poverty, deforestation, poor land management and rapid urbanisation.

Advice on drawing your Philippines map

- Practise drawing the outline of the islands (hard!)
- Add the earthquake risk (whole country, but concentrated near plate boundary).
- Indicate areas at greatest risk from typhoons. Essentially the south is less at risk and the northeast at greatest risk. Note the season/months.
- Mark on and name key volcanoes, e.g. Pinatubo, Taal and Mayon (17 are active).
- Mark on Manila (megacity).
- Annotate central highland area as a site of landslides, e.g. Leyte.
- Annotate small islands as site of local tsunami risk, e.g. Biltran.

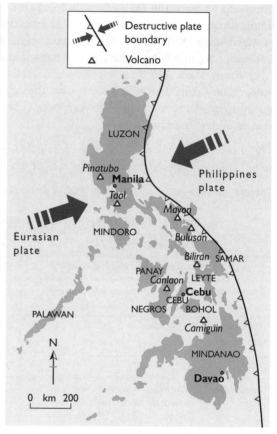

Figure 12 Sketch map of the Philippines

Table 2 shows the types of disasters the country experiences.

Table 2 Disasters in the Philippines, 1905–2007

	Number of events	Total killed	Total injured	Homeless total	Affected total	Damage ($m)	Example
Drought	6	8	0	0	6 million	64,000	April 1998, 2.5 million severely affected
Earthquake	21	9,580	13,051	3,985	2.25 million	844,485	Manila (1990) 6,000 killed
Flood	72	2,716	570	500,000	11.25 million	446,361	July 1972 2.7 million affected
Slide	25	2,604	381	23,000	310,663	12,258	February 2006 1,126 killed
Volcano	20	2,996	1,188	79,000	1.5 million	23,961	Taal (July 1911) 1,335 killed Pinatubo (July 1991) 700 killed
Tsunami	5	69	0	0	5,250	6,000	Worst tsunami in 1976
Typhoon	241	35,983	29,178	6.25 million	86 million	9,018,574	November 1991 6,000 killed

Compulsory case study: the California coast

Figure 13 summarises the risks and the links between them.

- Earthquakes are the main risk. Large shallow earthquakes occur along the swarm of faults associated with the San Andreas fault — a conservative plate boundary.
- River floods occur in El Niño years, and droughts and wildfires in La Niña years.
- Fogs occur in the San Francisco Bay area.
- Landslides are a frequent secondary hazard from floods or earthquakes.
- Coping capacity is high in this wealthy area, with much high-tech disaster preparedness.
- Economic costs of major disasters are high especially if disaster strikes in a megacity.

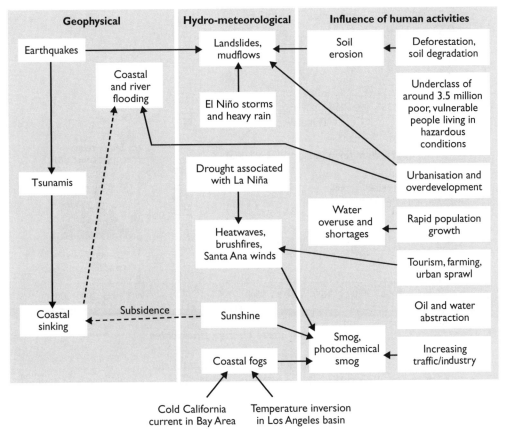

Figure 13 Disaster web for the California coast

Points to consider about multiple hazard zones

What is the effect of being a multiple hazard/disaster hotspot?

1 The hazards often occur as part of a complex web, making their impact greater and more complex to manage.

2 The frequency of events does not always make the impact easier to manage, although in wealthy countries systems can be put in place (contrast California with the Philippines, which only has good systems for geophysical hazards).

3 The magnitude of the hazard event, and the human geography of the area in which it occurs, are still probably the most important factors.

Climate change and its causes

Putting short-term climate change (global warming) in context

Climate change occurs at a range of different timescales, as shown in Figure 14.

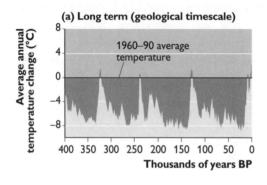

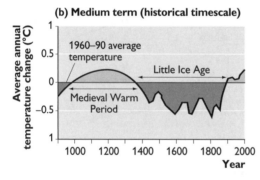

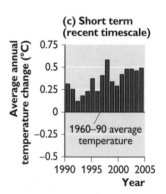

Figure 14 Climate change timescales

Climate change timescales and how the evidence is gathered

- **Long-term (geological)** — this essentially refers to the cycle of ice ages, which began some 450,000 years ago. Ice ages and the warmer periods in between them, known as **interglacials**, occurred at almost regular intervals throughout the geological period known as the Quaternary period.

The best evidence for long-term climate change comes from ice cores from the Greenland and Antarctic ice sheets. Air can be sampled from bubbles trapped in the ice as it was laid down over thousands of years, and carbon dioxide levels in the former atmosphere can be reconstructed. During the ice ages carbon dioxide levels are very low (180 ppm) but in the warm interglacials they rise to 280 ppm. (Present-day levels are around 384 ppm.) Further evidence is provided by pollen extracted from sediment cores from peat bogs. Pollen grains are preserved in waterlogged peat. Different-shaped pollen grains signify different trees, some of which lived in Arctic conditions (birch) and others in warmer conditions. Unfortunately, the evidence is incomplete and long sequences are rare.

- **Medium-term (historical)** For the period between AD 1000 and 1850, there is no direct climate evidence so scientists rely on a series of proxy records to reconstruct the climate. For example, evidence from paintings, poems, diaries etc. suggests that there was a 'Little Ice Age' from around 1500 to 1800 during which the Thames froze over regularly. The pattern of retreating glaciers can also be seen from paintings of the Alps, suggesting that their most recent maximum extent was around 1850. Records of dates of French grape harvests have been used to identify warm and cold periods, and the thickness of tree rings is also an excellent indicator. In higher temperatures, with more sunlight and abundant precipitation, trees grow well and have thicker tree rings, whereas in colder periods their growth is stunted. For some trees, such as the ancient bristlecone pines in western USA, evidence goes back 4,500 years.
- **Short-term (recent)** is much more straightforward to analyse. Quality instrumental records have existed for the last 100 years, as have detailed records of the response of ice sheets and glaciers. The ice sheets of Greenland, the Canadian Arctic and Antarctica are all aerially surveyed and monitored continuously. Air and ocean temperatures are recorded and ecosystem changes are monitored because their disappearance would have such an impact on world systems.

The drivers of climate change

Natural causes are the main drivers of long-term and medium-term climate change.

It is generally accepted that **Milankovitch cycles** (orbital eccentricity, axial precision and axial tilt) are the basic cause of **long-term climate change**. These are variations in the Earth's orbit around the sun that cause a change in the amount, distribution and seasonal timing of solar radiation. The theory is named '**astronomical forcing**'. In support of the theory is the regular pattern of ice ages and interglacials. The actual impact of these orbital changes is enough to change global temperatures by 0.5°C, whereas records show that ice ages were about 5°C colder than the interglacials. The cycles may have been just enough to trigger these more substantial climate changes. For example, feedback mechanisms (such as greater reflection of solar radiation from increased snow cover that has a higher albedo) could tip the Earth into an ice age.

Medium-term climate change in historical times, which includes the Medieval Warm Period and the Little Ice Age, has been linked to changes in **solar output** at timescales longer than the regular 11-year sunspot cycle. Sunspots (dark spots on the sun's surface caused by intense magnetic storms) have been well recorded over the last 400 years. It would seem that the Little Ice Age could be linked to a long period with almost no sunspot activity.

Volcanic activity at a super-volcano scale can also alter global climate. Volcanoes eject huge volumes of ash, water vapour, sulphur dioxide and carbon dioxide into the atmosphere, which reduces the amount of sunlight received at the Earth's surface. However, even for a very large eruption such as that of Krakatau in the nineteenth century, the temperature changes are short-lived (1–2 years). Patchy historic records do suggest increased volcanic activity during the Little Ice Age but this is unlikely to be its cause.

Short-term climate change could be part of the longer-term cycle of natural causes. However, scientific opinion increasingly supports the view that the unprecedented scale and rate of change is linked to the huge rise in **greenhouse gas emissions** and is therefore a direct result of the enhanced greenhouse effect (**atmospheric forcing**). Concentrations of carbon dioxide, ozone, methane, nitrous oxide and chlorofluoro-carbons in the atmosphere have grown significantly since pre-industrial times as a result of human activity. These contribute to the enhanced greenhouse effect (Figure 15).

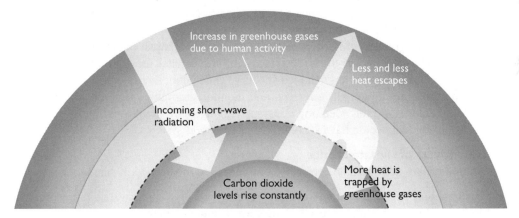

Figure 15 The enhanced greenhouse effect

Eight key questions on climate change for you to consider

1 Is short-term climate change part of the longer-term cycle of climate change or is it happening at an unprecedented rate and therefore something special?
2 Is this short-term climate change caused by natural or human (anthropogenic) causes?
3 What is causing global warming?
4 Is there any interconnection between global warming and oscillations such as the El Niño Southern Oscillation (ENSO)?
5 Why are the impacts of climate change so difficult to predict?
6 How and why do they differ from place to place and how are they likely to change over time?
7 What makes climate change the world's most difficult problem to solve? Why does this cause conflict?
8 Will we reach the tipping point or come back from the brink?

The impacts of global warming

Direct impacts

The two compulsory case studies emphasise the scale of the environmental, social and economic impacts of global warming. The **Arctic** has been selected because the environmental impacts are happening here and now. It is both a 'barometer' of environmental impacts and an early warning as to what the rest of the world might face. In contrast the continent of **Africa** is only expected to have average rises of 2–3°C by 2100, but the social and economic impacts are made far worse by the sensitivity of the environment, the vulnerability of the people and their low adaptive capacity to issues such as drought, water shortages, food insecurity, worsening health, debt and endemic poverty.

Compulsory case study: environmental and ecological impacts of climate change on the Arctic region

In the past few decades, average Arctic temperatures have risen at twice the rate of the rest of the world (3–4°C in the last 50 years in Alaska and northwest Canada). Over the next 100 years they could rise a further 3–5°C over land and up to 7°C over the oceans. This is already leading to the melting of the Greenland ice sheet (see Figure 16). The Arctic Impact Assessment website (**www.acia.uaf.edu**) provides a detailed assessment.

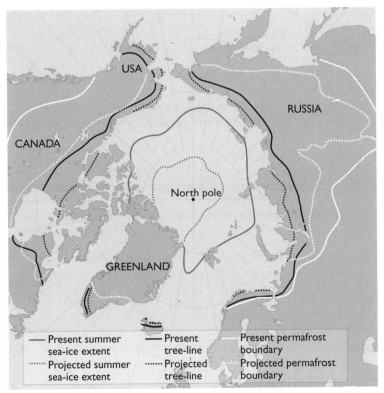

Figure 16 Arctic region: summary of key changes

Impacts on natural systems
- **Vegetation shifts.** Vegetation zones are predicted to shift northwards. This shift will destabilise existing food webs. The longer, warmer growing season will benefit Arctic agriculture although soils will be a limiting factor.
- **Thawing permafrost.** Up to 40% of total permafrost is expected to thaw, especially in Siberia. In some areas, lakes and rivers will drain. This will have an impact on species, particularly freshwater fish such as the Arctic char and lake trout.
- **Increasing fires and insects.** Global warming will increase forest fires and insect-caused tree death, which may have an impact on old-growth forest, a valuable habitat.
- **Ultraviolet impacts.** Increased ultraviolet (UV) radiation will reach the Earth's surface as snow and ice cover is lost. UV radiation destroys phytoplankton at the base of the marine food chain.
- **Carbon cycle changes.** The replacement of Arctic vegetation with more forests will lead to higher primary productivity and increased carbon dioxide uptake, but increased methane emissions could counterbalance this positive impact.
- **Other impacts.** Increased coastal erosion as thawing permafrost weakens the coast, and there are more waves and storm surges as the protection of sea ice is lost.

Impacts on animal species
- Species will shift north with forests. Some species are likely to suffer major decline.
- Marine species dependent on sea ice, including polar bears and ice-living seals, will decline. Birds like geese will have different migration patterns.
- Land species adapted to the Arctic climate, including Arctic fox and caribou, are at risk.

Impacts on society
The ecological and environmental changes described above will mean:
- loss of hunting culture and decline of food security for indigenous peoples
- need for herd animals to change their migration routes
- enhanced agriculture and forestry
- the Arctic will become more accessible, and vulnerable to exploitation for oil, gas, fish and other resources

Compulsory case study: socioeconomic impacts of climate change on the continent of Africa

Africa is the continent that makes the least contribution to global warming, yet it is the most vulnerable to climate change. Much of its population is dependent on climate-sensitive resources such as local water and ecosystems, and has a limited ability to respond to changing climate because of poverty.

It is predicted that temperatures in Africa overall will rise by 3–4°C by 2100. Rainfall is likely to increase in the equatorial region but decrease to the north and south of that band.

Figure 17 summarises Africa's vulnerability to global warming.

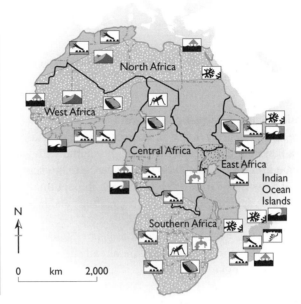

Figure 17 Global warming impacts in Africa

- **Water issues.** Life in Africa is regulated by access to water for agriculture, domestic use and hydroelectric power. Demand outstrips supply of water for 25% of Africans. However, enough water is available in most parts of Africa. Poverty is the key reason why millions have no access to safe and reliable water supplies. Water stress could lead to wars, global migration and famine.
- **Food insecurity.** Seventy per cent of the population are subsistence farmers, many of whom will not be able to feed themselves should water supplies dry up, pasture quality deteriorate or crops fails. Increased locust plagues may also threaten food supplies.
- **Natural resources.** Poor people, especially those living in marginal environments, depend directly on wild plants and animals to support their way of life. Loss of biodiversity due to climate change will threaten them.
- **Health.** Vector-borne diseases (e.g. malaria) and water-borne diseases (e.g. diarrhoea) could increase with climate change, overloading inadequate healthcare systems.
- **Development of coastal zones.** Movement of environmental refugees from the countryside puts pressure on the coastal zones; 60% of Africans live in coastal zones, many of which are at risk of coastal erosion and flooding. The threat from these is likely to increase as a result of rising sea level. If the coastal zones were flooded, much of the continent's infrastructure of roads, bridges and buildings would also be lost.
- **Poverty.** At the root of Africa's vulnerability is poverty. The problem is made worse by conflicts (e.g. Darfur). An unjust trading system forces many countries to sell their exports at a low price. Above all, the burden of unpayable debt means that no money is available for the mitigation of climate impacts and the introduction of adaptive strategies.

Exam tip

In the next section (on coping with climate change), a number of solutions are outlined for the continent of Africa. Beware of 'total doom and gloom'. The following websites are useful for reference: **www.iied.org** and **www.oxfam.org.uk.**

Indirect impact: rising sea levels

Predicting sea-level rise is complex and uncertain. It is caused by both thermal expansion of the oceans (increase in volume with warming) and melting ice sheets. Some areas are particularly vulnerable to rising sea levels:

- the world's largest deltas, e.g. Brahmaputra/Ganges in Bangladesh, Nile and Mississippi; these have large populations and high risk of exposure to storm surges, flooding and sea-level rise
- small low-lying islands (coral atolls in the Indian and Pacific Oceans)
- areas close to sea level that are already heavily defended, e.g. the Netherlands and eastern England (the Wash)
- places such as Hong Kong and Singapore, where new developments have been built on reclaimed coastal land

The first environmental refugees have already been registered from the Chra Islands of Bangladesh and Tegua in the South Seas.

Exam tip

Many of you will be studying 'Crowded coasts' as an option in Unit 2, so you can use some examples you have studied for that. For the Unit 1 exam, you will need several short examples looking at the impact of rising sea levels on different coastlines.

Predicting emissions and their likely impact

The Intergovernmental Panel on Climate Change (IPCC), first formed in 1988, collects data on greenhouse gas concentrations, sea levels and ice stores. As the prediction of climate change is so uncertain, the IPCC envisages different possibilities based on a range of emissions scenarios (Figure 18).

- **High emissions scenario:** rapid economic growth, increasing populations, reliance on fossil fuels and business as usual.
- **Medium-high emissions scenario:** more self-reliance, increasing populations and economic growth.
- **Medium-low emissions scenario:** population growth slows, clean and efficient technology, reduction in use of fossil fuels.
- **Low emissions scenario:** local solutions to sustainability, slower rate of population increase, less rapid technological change.

Why the uncertainty?

Predictions of emissions levels and their impacts are difficult, because it is hard to predict the following:

- the level and nature of economic development, particularly in countries like India and China, which will determine greenhouse gas (GHG) emissions
- what degree of international action will be taken to reduce emissions

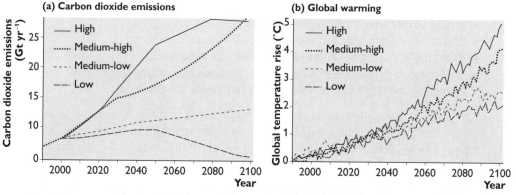

Figure 18 Climate change predictions by the IPCC

- the inertia in the system — even if GHG emissions stabilise, climate change will continue
- the impact of positive feedback, for example as permafrost areas thaw due to global warming the powerful GHG methane will be released, increasing warming still further

Too late to act?

The scientific evidence points to increasing risk of serious, irreversible impacts from climate change. The point when change becomes irreversible and catastrophic is called the **tipping point**.

The current level (2007) of greenhouse gases in the atmosphere is nearly 400 ppm carbon dioxide equivalent. Even if the annual flow of emissions did not increase beyond today's rate, the stock of greenhouse gases in the atmosphere could reach 550 ppm carbon dioxide equivalent by 2050. However, the annual flow of emissions is accelerating as fast-growing economies invest in a high-carbon infrastructure and as demand for energy and transport increases around the world. The level of 550 ppm could be reached as early as 2035. At this level there is a strong chance of a global average temperature rise exceeding 2°C, which many regard as the tipping point for temperatures. Remember: the Arctic is already showing this profile of accelerating change.

Under a business-as-usual scenario with no serious attempt at mitigating the impacts, greenhouse gases could more than treble by 2100, giving at least a 50% risk of exceeding a 5°C global average temperature change in the following decades. This huge rise would take humans into unknown territory — disaster on an unimaginable scale.

Coping with climate change

The role of mitigation and adaptation strategies

Mitigation involves reducing the output of greenhouse gases and increasing the size of greenhouse gas sinks. It is a longer-term solution.

Examples of mitigation strategies include:
- setting targets to reduce carbon dioxide emissions
- developing energy-efficient strategies in all economic sectors
- switching to renewable energy sources such as wind power
- 'capturing' carbon emissions from power stations and storing them
- enhancing the size of carbon sinks, for example by afforestation

Arguments for mitigation are strong as it is necessary to avoid irreversible climate change and mitigation is the ultimate long-term solution. For mitigation to work, it has to take place at all scales, from local to global. Figure 19 shows the scales of mitigation and the key players.

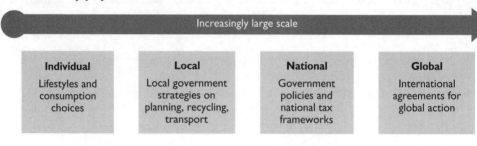

Figure 19 Scales of mitigation

Adaptation means changing our lifestyles and businesses to develop strategies to cope with the increasing dangers from climate change — rising temperatures, more extreme weather, increasing water, food and health insecurity, and rising sea levels. It is an immediate solution.

Examples of adaptation strategies include:
- coastal defences or managed retreat of a coastline vulnerable to sea-level rise
- developing a second generation of drought-resistant GM crops
- enlarging conservation zones or developing wildlife corridors to allow for shifting ecosystems and habitats

There are strong arguments for adaptation (i.e. climate proofing). Many scientists argue that the effects of climate change are happening already (as in the Arctic) and even if humans stopped polluting the atmosphere now climate change would continue. Adaptation and mitigation go hand in hand and both are needed.

Exam tip

For further details of mitigation and adaptation strategies, explore the IPCC website.

Key players in managing climate change
Kyoto Protocol: governments reaching global agreement?
At the 1992 Rio Earth Summit, the UN Convention Framework on Climate Change was agreed. It aimed 'to achieve stabilisation of greenhouse gas concentrations at a low enough level to prevent dangerous anthropogenic interference with the world's climate systems'. It led to the agreement of the Kyoto Protocol in 1997.

However, putting this framework into practice led to extremely complex and ongoing negotiations. Many countries delayed signing the treaty because they felt cutting emissions would damage their economies. Russia did not sign until 2005 and Australia, the second largest per capita polluter, did not sign until 2007. The USA, which produces 25% of global emissions, initially signed but withdrew in 2002 following the election of George Bush.

Complex systems were introduced allowing 'trading' of carbon credits: that is, buying unused emissions from other countries or businesses. Carbon sinks, such as planting forests, were allowed, so that countries can 'offset' emissions. Critics argue that both of these systems allow polluters to continue to pollute.

Overall reductions in emissions have been under 1%, largely because of increasing emissions from countries such as China, India and Brazil, which were not obliged to cut emissions.

Big businesses
Some transnational corporations (TNCs) are bigger economically than developing countries and their actions wield enormous economic power. Initially, big oil and car companies were opposed to cutting emissions, arguing that reducing pollution would cost money, profits and jobs. Now, however, leading TNCs produce annual environmental reports and many are committed to greener growth. For example, General Electric, the world's largest company, has committed itself to double its research and development spending on clean technologies, double its revenue from renewables and energy efficient products and reduce its greenhouse gas emissions by 30% by 2012. This change of attitude can be explained by pressures from customers and investors and the need to display a good public image. There are big profits to be made in renewables and new clean technology. Incentives are available for offsetting schemes and clean development mechanisms, and governments are increasingly taxing carbon emissions.

National governments
National governments play a vital role in developing strategies to reduce emissions, by:
- promoting energy-efficient technology, e.g. ecobuildings, green transport
- encouraging energy generation from nuclear power (France), renewables (Scandinavia and UK) or biofuels (USA) instead of fossil fuels
- promoting carbon storage by sequestration of power-station emissions (Norway) and creation of sinks by tree planting
- awarding huge grants for developing carbon neutral ecotowns
- legislation and fiscal policies, e.g. taxing pollution

Local governments
Local governments tend to reflect the strategies of their national governments. In the UK, London has led the way with legislation such as the congestion charge. Leicester is the UK's first environmental city. Figure 20 shows a range of emission-reducing schemes available to local authorities.

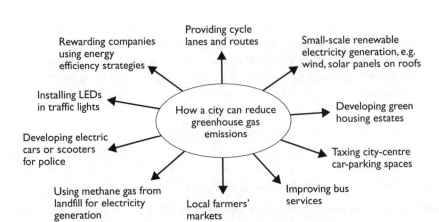

Figure 20 How a city can reduce greenhouse gas emissions

Areas throughout the world have developed Local Agenda 21, which emerged from the 1992 Rio Earth Summit. It encourages local authorities to implement a raft of sustainable strategies to improve the environment and reduce carbon emissions. Many people are encouraged by positive local actions on recycling, green transport and small-scale schemes (e.g. combined heat and power) to do their bit for climate change.

NGOs

Most environmental NGOs operate in advocacy and education roles to bring pressures on governments and individuals to act. They campaign on the causes, impacts and solutions to climate change for a number of environmental and socio-political reasons.

> **Exam tip**
>
> Look at the websites of some NGOs which campaign on climate change (e.g. Camp for Climate Action, Eco Equity). Find out what their role is and how they campaign. Assess the degree of bias in their information.

Individuals

The key to getting individuals involved in cutting emissions is education. Calculating the **carbon footprint** (the amount of carbon dioxide emitted by an individual or a family or school each year) is a very good personal check on performance. It creates an opportunity to reduce the footprint by changing patterns of consumption and transport use, by installing energy-efficient systems and, if all else fails, by offsetting carbon-emitting actions. However, many of these changes may be challenging, sometimes even impossible to implement. Think about some of the following ideas:

- walking, cycling or taking the bus/train to work or college, i.e. not using the car
- buying locally produced food — often at higher prices and with limited choice
- switching energy supplies to renewables — this is costly
- becoming energy efficient in the home, e.g. not leaving the television on standby, not using the tumble dryer in summer

The challenge of global hazards for the future

The enormity of the challenge

Global warming and oscillations like El Niño and La Niña are causing rising numbers of hydro-meteorological hazards (wind storms, drought and floods). The increasing numbers of vulnerable people, who have low adaptive capacity, means that these hazards frequently develop into disasters.

Water shortages

Climate change will exacerbate water stress (lack of a safe, reliable supply), which is currently experienced by some 40% of the world's people. As a result of rising temperatures, greater incidence of drought or loss of supplies from glaciers, around half a billion more people will experience a physical shortage of water, i.e. demand exceeding supply. As poverty levels rise, there may be many more people not able to afford clean water, i.e. an economic shortage of water. Water is needed not only by people themselves but also for their rainfed agriculture and stock. Water shortage will have a profound effect on food security.

Food insecurity

Food insecurity means people's inability to access an adequate diet. Climate change will exacerbate famine in parts of Africa, South America and south Asia, as an increase in extreme weather events leads to falling yields in cereals. Rising poverty will mean that fewer people can afford to access food and more will rely on food aid. Climate change has not only led to reductions in crop yields. The desire to cut emissions has also led to the planting of millions of hectares of maize/soy beans/sugar cane for **biofuels** at the expense of growing cereals for food.

The relationship between global warming and food security is complex. The growing of cash crops (notably tea) in some regions will be affected by changing conditions, but elsewhere there will be some benefits, such as a longer growing season in the Arctic and more rainfall in Kenya.

In theory, agriculture should be able to adapt to climate change using a variety of technologies, from intermediate (appropriate) technology such as soil conservation schemes in Mali, to high technology such as the long-awaited second generation of drought-resistant/salt-tolerant GM crops. The reality is that ecosystems are extremely sensitive, the vulnerability of people high, and the adaptive capacity limited in the developing world.

Tackling the challenge

A number of sustainable strategies can be used to tackle climate change. However, many of them have costs as well as benefits. Green strategies include a range of so-called 'eco-friendly' projects.

Tree planting

Trees act as a 'sink' by taking in carbon dioxide and 'fixing' it in the form of hydrocarbons (plant matter). However, the benefits are not felt for at least 10 years as a growing tree releases more carbon dioxide than it absorbs, especially if the ground

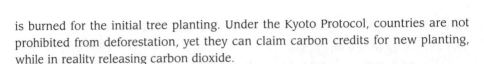

is burned for the initial tree planting. Under the Kyoto Protocol, countries are not prohibited from deforestation, yet they can claim carbon credits for new planting, while in reality releasing carbon dioxide.

Renewable energy projects

These cut down on the use of fossil fuels. Some small-scale projects — such as solar panels, microhydro turbines, biogas converters and wind turbines — are of vital importance to remote areas in developing countries, revolutionising the quality of life of the rural poor.

Any large-scale schemes, such as the Three Gorges dam in China or the new mega dam proposed for the River Congo, have all kinds of environmental and ecological negatives, as well as the obvious benefits of providing large quantities of 'green' electricity along with flood control, irrigation and tourism. In some countries that depend on hydroelectric power (HEP), there is concern that climate change will render HEP schemes useless because of drought affecting water supplies.

The most contentious development is probably that of **biofuels**. There are two groups of the so-called first generation of biofuels: **biodiesel** is usually made from soya beans or rapeseed, which is added to conventional diesel; **bioethanol** is made from corn (USA) and sugar cane (Brazil). There are huge concerns about the environmental damage and impacts on biodiversity resulting from destruction of rainforests or draining of wetland areas to grow these biofuel crops in countries such as Brazil or Indonesia. They also exacerbate the food-security crisis where they are grown instead of food crops such as wheat or rice. The question has to be asked whether the costs of growing biofuel crops outweigh the benefits and whether it is a truly 'green' fuel. In the USA, the main driver seems to be combating the rising price of oil rather than reducing greenhouse gas emissions.

Wind farms produce green electricity but large areas of turbines are required to generate comparatively modest amounts of power. Some of the best sites in terms of wind occur in areas of outstanding natural beauty and development of wind farms is hotly opposed in such locations. Recently a large scheme in the peatlands of the Isle of Lewis in the Outer Hebrides was rejected on environmental grounds, with 11,000 residents signing a petition against it.

A major problem with renewable sources of energy is that they tend to be more expensive than fossil fuels. Nuclear power is another option, but opinion is sharply divided as to whether it is a *green* option, because of waste and safety issues.

Community-based strategies

These often work well as they are developed in a 'bottom-up' fashion by the community, as opposed to being forced upon them by higher authorities — that is, they are proactive. Communities in countries at all stages of development can take the initiative in producing schemes that mitigate or adapt to climate change. The only problem about bottom-up schemes is that they are often locally based and not easily replicated everywhere. Their success also depends on the energy of a 'hard core' of leaders and organisers to motivate the community.

Examples include the creation of low carbon communities in towns and villages such as Wolvercote (Oxfordshire), and polythene-bag-free towns such as Modbury (Devon). There are numerous community-based initiatives, frequently financed by NGOs, in developing countries. These are known as **capacity building schemes**, which teach local people to adapt to climate change. Community forest projects are also widespread.

Energy efficiency
This is possibly the most effective way forward, although initial costs of any scheme are high. Long-term energy efficiency not only reduces emissions, but it actually cuts costs and improves local pollution levels. Avoiding dirty development in the first place is clearly the way ahead for developing countries.

Methods of increasing energy efficiency include:
- remodelled factories with clean, efficient industrial processes and optimum use of energy
- redesigned houses with modern boiler systems and full insulation
- greener and more efficient power stations
- green transport using new, greener fuels (hybrid technology or hydrogen-powered)
- use of recycling, e.g. waste heat from incineration for central heating systems

One word of warning: India and China, because of their huge size (their combined populations comprise one-third of the world's people, 2.5 billion) and rapid economic growth (average of 6–10% per year) have a critical role to play in controlling emissions. Even if all the more developed countries cut their emissions drastically the following statistics suggest that the economic growth of China and India will lead to more emissions:
- The increase in China's greenhouse gas emissions from 2000 to 2030 will nearly equal the increase from the entire industrialised world. In 2008 China became the world's number one greenhouse gas emitter.
- India's greenhouse gas emissions are estimated to rise by 70% by 2025, with India's energy consumption rising even faster than China's.
- Both India and China are desperate for energy to fuel their economic expansion, in order to bring their citizens out of poverty and improve the quality of their lives — 40% of India's population currently lack regular access to electricity. Environmental concerns inevitably lag behind the need for growth, but both countries are concerned about pollution and energy shortages.

Solutions in a more hazardous world
The Hyogo Framework for Action (Figure 21) was developed in 2005 by the World Conference on Disaster Reduction held in Kobe, Japan. At its heart is risk reduction by building resilience to disasters and by overcoming the underlying factors that lead to vulnerability, notably poverty. Their aim is to exploit the Kyoto mechanisms, such as the Special Climate Change Fund and the Clean Development Mechanism, to allow developed countries to pay for greenhouse gas cutting projects in return for carbon credits.

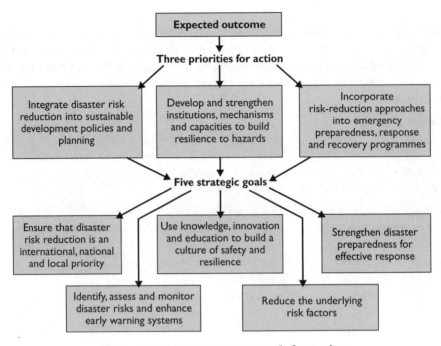

Figure 21 The Hyogo Framework for Action

The framework can be applied at all scales, from local to international. In Unit 2 you will be investigating recurring hazard events — either river flooding or the impact of coastal storm surges. At an international scale, a mega-disaster such as the 2004 south Asia tsunami can be researched, looking at how the new coastal zoning and coastal economic development schemes, and the Indian-built tsunami early warning systems, have fulfilled the three priorities for action in the Hyogo Framework. At a regional scale you could research poverty in sub-Saharan Africa to see how the various countries are using the framework and at a local scale you could look at flooding and the management of flood risk (using the Environment Agency website: **www.environment-agency.gov.uk**).

In conclusion, Figure 22 summaries the importance of an integrated approach to management of the rising risk from climate change and hydro-meteorological hazards.

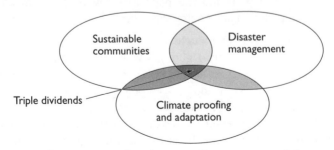

Figure 22 The importance of an integrated response to risk management

Going global

'Going global' examines the process of globalisation: what drives it and its winners and losers. Increasingly rapid social and economic change has accompanied globalisation, altering family structures, work patterns and jobs. Increasing mobility has encouraged migration on an unprecedented scale, especially to cities. 'Going global' explores the benefits and costs of an increasingly global world and examines some of the challenges and potential solutions. Figure 23 shows how the topic content links together.

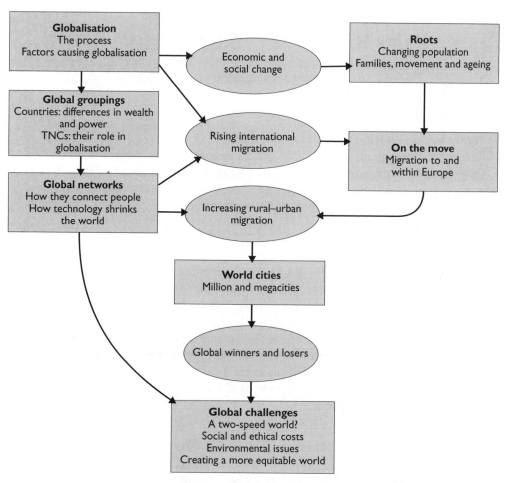

Figure 23 Topic concept map

A key theme of globalisation is the quickening pace of technological change. Figure 24 contrasts the gradual adoption of older technologies with the extraordinarily rapid growth of mobile phones and the internet. The first UK text message was sent in 1992 and broadband only became widely available in 2000.

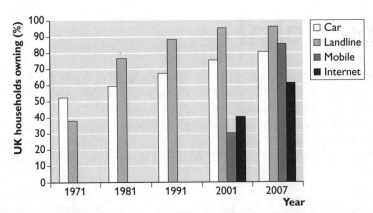

Figure 24 Changing UK household access to four technologies, 1971–2007

Life without broadband, mobile phones, iPods and social networking sites might be hard to imagine, but only 10–15 years ago none of these existed.

Globalisation

What is globalisation?

The International Monetary Fund (IMF) defines globalisation as 'the growing economic interdependence of countries worldwide through increasing volume and variety of cross-border transactions in goods and services, freer international capital flows, and more rapid and widespread diffusion of technology'.

Interdependence suggests that in a globalised world, events in any one country will quickly spread to others.

Economists argue that globalisation is a centuries-old process. In the nineteenth century the British empire was a 'global network' connected by:

- a network of undersea telegraph cables, connecting to overland telegraph, allowing for rapid communications
- extensive trade in raw materials from colonies and manufactured goods from the UK
- a shared head of state (Queen Victoria for most of the nineteenth century)
- global military power, especially the Royal Navy
- attempts at imposing a shared culture (international cricket is one of the clearest links back to imperial times)

Types of globalisation

Globalisation takes several forms, and all involve increasing interdependence and interconnectedness:

- **Economic globalisation** including the growth of transnational corporations (TNCs), with a global presence and brand image, the global spread of foreign direct investment (FDI) and growth in world trade.
- As TNCs have grown, **cultural globalisation** has followed. McDonald's, present in 120 countries with 30,000 outlets, ensures that increasingly people eat similar food, and CNN that people hear similar news.

- **Political globalisation** is the dominance of Western democracies (G8 countries) in decision making and the view that democratic, consumerist societies are the 'model' to which others should aspire.
- **Demographic globalisation** results from increasing migration and mixing of populations.
- **Environmental globalisation** involves the realisation that global environmental threats require global solutions.

Factors that have accelerated globalisation

Many factors have contributed to our increasingly globalised world. Table 3 explains and evaluates the importance of these factors.

Table 3 Factors encouraging globalisation

Factor	Explanation
Free trade	Removing trade tariffs and quotas has promoted easier and faster trade.
International organisations	The World Trade Organization has promoted free trade, and trade blocs have encouraged free trade within and between groups of countries.
Oil money	High oil prices in the 1970s created wealth in OPEC countries. Money was loaned to developing nations, and this 'kick started' their industrialisation.
TNCs	TNCs have shifted production to the developing world and created global connections and trade links.
Communications technology	Satellite and fibre optic communications have led to growth of mobile and internet communications, and falling communication costs.
Transport technology	Containerisation since the 1960s has led to increasingly cheap, automated and efficient methods of transport. Ships can carry 9,000+ containers. For high-value, low-volume goods, and people, cheap air travel has revolutionised transport.
Financial deregulation	Government controls on banks, currencies, interest rates and companies have all decreased. This makes investment easier and profits higher.
Consumers	The global consumer has contributed to soaring demand for goods from all corners of the world.
The media	Large global media corporations have global reach and present a similar 'world view' of the news, contributing to the sense of a connected world.

Opportunities

Globalisation presents countries, companies and individuals with numerous opportunities. Great wealth has been created through trade. China's exports rose from $200 billion in 2000 to over $1,000 billion in 2007. Foreign direct investment has led to industrialisation, resource exploitation and property development. This creates jobs and wealth, plus a return for the investor. FDI into China increased from $4 billion in 1991 to $64 billion in 2006. Individuals have benefited through migration to economic hotspots such as Dubai and Bangalore. Developed world economies are increasingly knowledge driven — information is exchanged, rather than goods. Investment in education is one way of increasing opportunities, another is 'poaching' key people, increasingly from all corners of the globe.

While many people welcome globalisation, it does have drawbacks:

- It has led to growing inequality — the rich have grown richer, and the poorest relatively poorer.
- It is a process dominated by TNCs and governments, and ordinary people may feel like 'pawns' (e.g. lack of unions, loss of jobs due to global shifts and exploitation of workers in developing nations).
- It encourages unsustainable economic growth, excessive resource and energy use, and therefore has negative environmental consequences.
- It tends towards a 'Westernised' global culture where local traditions, languages, food and art disappear.

Population movements

Globalisation causes migration. Most migrants move for economic opportunity. The movements may be legal or illegal. As some parts of the world have become wealthier, the gap between the wealthy and the poorest has widened. This encourages migration for greater opportunities. The world migrant stock in 2006 stood at:

- 130 million legal economic migrants (80 million in 1970)
- 30–40 million illegal migrants (largely economic)
- 8–10 million refugees
- 25 million internally displaced people

When migrants cannot move legally, they move illegally. These movements are towards economic hotspots and are usually from the developing to the developed world:

- the USA — major flows from Mexico, including up to 10 million illegal immigrants across the US–Mexico land border
- the EU — including up to 5 million illegal migrants across the southern EU fringe in Spain, Malta and Italy
- the middle east — attracts large numbers of migrants from south Asia to work in construction and domestic service

Global groupings

Rich, developed countries have retained their dominance during the era of globalisation. A lucky few countries have joined this rich group, for example the 'Asian Tigers' of Singapore and South Korea, and more recently China. Economic power is concentrated in the three global economic cores of North America, Europe and Asia. Of the 500 largest TNCs in the world, 162 are based in the USA and 67 in Japan. The three cores are referred to as the **triad**. Around 80% of all global economic wealth and trade is concentrated in this triad, which is linked through a complex system of global finance, stock exchanges, international airports and government centres.

Economic groupings

Geographers classify countries by their level of economic development, specifically their **GNI** (gross national income). Dividing the world into the 'rich North' and 'poor South' is a common device — but the world is more complex than this. Figure 25 shows that between 1975 and 2002 some groupings of countries became wealthier, but others stagnated or became poorer. This has created a world of the rich, those getting richer, and those either stagnating or getting poorer.

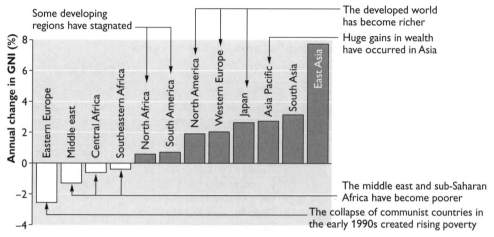

Figure 25 Change in GNI, 1975–2002

Economic groupings are useful in geography because they allow us to compare different development pathways. Countries can, and do, move between groups. Singapore was once classed as an NIC but is now a fully developed country. Table 4 outlines the most commonly used economic groupings.

Table 4 Economic groupings of countries

Economic grouping	Examples	World Bank groupings by annual per capita income, 2006	Economic structure
More economically developed countries	UK, USA, Australia, Greece, Singapore	High income ($11,115 or more)	Services account for 70%+ jobs. TNCs are based in these countries; research and development and quaternary industry important.
Newly industrialised countries	South Korea, Brazil, Taiwan, Chile, China	Upper middle income ($3,596–11,115)	Service economies with significant manufacturing; primary and secondary industries of decreasing importance.
Recently industrialised countries	Thailand, Indonesia, Tunisia	Lower middle income ($906–3,595)	Significant primary employment. In key cities, manufacturing is growing rapidly.
Less economically developed countries	Egypt, Peru	Low income (under $905)	Primary industry provides 40%+ of jobs; services and manufacturing developing. Development is slow.
Least developed countries	Malawi, Bangladesh, Haiti		Dependent on primary industry. High levels of poverty persist. Many LDCs are getting poorer in absolute terms.

Political groupings

Formal groupings of countries, called inter-governmental organisations (IGOs), exist where countries have signed agreements for economic gain. Such groups contain 'like-minded' countries usually at similar levels of development. In most cases the groups protect the interests of member states (Table 5).

Table 5 Political groupings of countries

Grouping	Members	Role	Global importance
European Union (EU)	*Formed in 1957* 27 European countries, including the UK	An economic union, with internal free trade and population movement	Accounts for only 7.5% of world population but 31% of the world's total GDP
Organisation for Economic Cooperation and Development (OECD)	*Formed in 1961* 30 members, 25 of which are fully developed economies. All are democratic market economies	The OECD works to ensure the rich developed economies run smoothly. It monitors economic performance and works to reduce corruption and bribery	75% of global GDP is generated within the 30 member countries
Organisation of the Petroleum Exporting Countries (OPEC)	*Formed in 1960* 12 major oil exporters in the middle east (e.g. Saudi Arabia, Kuwait), South America and Africa	OPEC works to safeguard the interests of the major oil exporting countries. It has a large influence on the global oil price	OPEC countries have 65% of global oil reserves and 35% of production
Group of Eight (G8)	*Formed in 1974* UK, USA, France, Canada, Germany, Italy, Japan, Russia	An informal forum, really a club of the super-rich and powerful countries	Accounts for 65% of global GDP, but only 15% of population
Group of 20 developing nations (G20)	*Formed in 2003* 21 countries, including China, India and Brazil	Formed specifically to press developed nations to open up their markets to developing world trade	Accounts for 60% of the world's population and 20% of GDP
Group of 77 (G77)	*Formed in 1964* Most African, Asian and Latin American nations (130 in total)	A loose grouping of developing nations formed to give a collective voice to the developing world	Its influence is lessening, especially as China is no longer a member

Trade blocs

Political groupings usually have economic motives, to press for the interests of their member states. More formal trade blocs, such as the European Union (EU), North American Free Trade Area (NAFTA) and Association of South East Asian Nations (ASEAN), allow free trade, without taxes, tariffs or quotas between member states. Countries wishing to export into the bloc often have to pay trade taxes and/or have quotas imposed. These external trade barriers protect trade within the bloc.

The World Trade Organization (WTO) works to reduce trade barriers and create free trade between blocs. Increasingly the world is caught between these two forces. Trade blocs have tended to become more common, while free-trade agreements have also increased. There is general consensus that free trade is good, but increasingly an unwillingness to make it more free. The current set of talks to increase free trade, the Doha Round, began in 2001, but agreement has not yet been reached.

International trade growth has led to some significant shifts in wealth and power:
- Developed nations have maintained their 'top slot'.
- The 'Asian Tiger' NICs have developed almost to developed-country levels.
- In the last 10 years, Brazil, Russia, India and China (the 'BRICs') have all gained economic power and wealth.
- Asian and Latin American NICs and RICs have grown but often in a 'boom and bust' fashion.
- Many African countries have barely benefited, with population growth outstripping economic growth, leading to income stagnation.

The role of TNCs

Transnational corporations (TNCs) are major companies with a global 'reach' and a presence (production, HQ, sales) in at least two countries. TNCs are economically powerful and politically influential, and are important creators of wealth. As Table 6 shows, the largest TNCs have turnovers (sales) equivalent in size to large countries. Some, such as Wal-Mart, employ enough people to populate an entire city. This gives them huge power. Investment decisions made by TNCs can be both a blessing and a curse.

Table 6 The world's five largest TNCs compared with some countries

World's five largest companies, 2007	Turnover in 2007 ($ billions)	Employees 2006–07	Equivalent country by total GDP ($bn) 2007
Exxon Mobil	390	81,000	Indonesia (410)
Wal-Mart Stores	374	1,700,000	Taiwan (375)
Royal Dutch Shell	355	112,000	Greece (356)
BP	265	104,000	South Africa (274)
General Motors	206	323,000	Portugal (219)

Positives of TNCs
- **Jobs:** the 200,000+ technology workers in Bangalore are employed by TNCs including IBM, Cisco Systems and Google.
- **Trade:** China's economic growth has resulted from TNCs locating manufacturing plants in its Free Trade Zones, boosting exports.
- **Connections:** TNCs' complex global networks create connections that tie local and national economies into the global economic system.

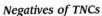

Negatives of TNCs

- **Exploitation:** TNCs have been accused of exploiting workers in the developing world by paying very low wages in 'sweatshop' conditions.
- **Unemployment:** Outsourcing jobs to the developing world can lead to job losses in developed countries.
- **Identity:** Local cultures and traditions can be eroded by TNC brands and Western ideas.

Costs and benefits of TNCs

Transnational corporations, with their ability to shift production from developed countries to new developing locations, have costs and benefits (Table 7). When TNCs invest in developing countries (foreign direct investment), they may also reduce production in their home country.

Table 7 Costs and benefits of TNCs

	Costs	Benefits
TNC source country (developed world)	Job losses due to outsourcing of jobs.	Environmental pollution from factories is 'exported' overseas.
	Abandoned production locations create derelict land.	TNC growth leads to higher profits, and more tax being paid.
	Outsourcing TNCs may become unpopular and suffer negative media coverage and falling sales.	Strong global companies are successful, securing HQ and research and development jobs.
TNC host country (developing world)	TNCs may pay no, or very low, taxes.	Economic growth due to job creation and rising consumption.
	New locations become polluted as environmental laws are weak.	Falling levels of poverty.
	Workers on low wages, with long hours, may be exploited and mistreated and suffer poor health.	Local supply chains may be created, leading to further job creation and business growth.
	TNCs may out-compete local suppliers, forcing them out of business.	TNCs demand infrastructure and communications that may benefit local people.

Exam tip

You need to know a case study of a TNC. Use the following headings for your research:
- Home country/HQ
- Market sectors in which the TNC is involved
- Employment and turnover data
- Global network: locations of production, sales and research and development
- Issues: you should use the internet to identify any environmental and social criticisms of your chosen TNC

Global networks

Global connections tie the world together as never before, through:
- mobile phones, even in remote areas

- instant personal and business internet communication
- low-cost jet air travel
- cable and satellite media channels
- favourite brands — Coca-Cola, Subway, Nike — available globally

It is easy to forget that this connectivity is not available to everyone. Living in the UK with an annual income of £15,000 might allow access to most of the connections, but approximately 6 billion people worldwide live on lower incomes. This disparity means that some places benefit from global networks and become 'switched on' while others remain poor and 'switched off'.

Switched on: air travel

The first commercial jetliner, the De Havilland Comet, could carry up to 100 passengers a distance of 5,000 km in the 1950s. The 2007 Airbus A380 can carry up to 850 people a distance of 15,000 km. Low-cost airlines such as easyJet and Ryanair have made air travel available to a wide market and are spreading to the developing world, e.g. India's Air Deccan launched in 2003.

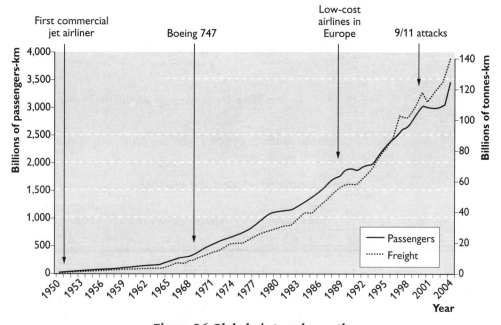

Figure 26 Global air travel growth

Increased air travel has a number of effects. It:
- reduces perceived distance between places
- increases cultural mixing, transforming exotic places into 'normal' experiences
- allows workers to move cheaply to areas of demand
- allows TNCs to move personnel and goods between production and sales locations

Air travel connects some places more than others, with the greatest connection density being between 'global hubs' in North America, Europe and Asia.

Switched on: cores and hubs

Global hubs are world cities or 'global regions'. All are in the economic core areas of the world and are centres of great wealth.

Hubs are extremely well connected to each other — for example, the financial, TNC and travel connections between London and Tokyo. Hubs act as regional nodes integrating the global economic cores such as the EU and connecting them to other cores. TNCs from hubs invest in emerging hubs, such as Bangalore in India and Shanghai in China, thus creating new connections.

Linkages between governments, TNCs and international organisations have produced a global super-rich class of media executives, lawyers, accountants, PR people, bankers and retailers who live in hubs and hop between them.

Switched on: the internet

Around 1.3 billion people were using the internet by 2007. This is extraordinary for a technology that was rare in the mid-1990s. However, growth has been uneven, creating a 'digital divide' between the developed and developing world. The divide links to income levels because the internet requires:

- a PC or laptop
- a complex and reliable telecommunications infrastructure
- internet service providers (ISPs)
- the ability to pay for a connection
- companies willing to invest in providing services, such as online booking and shopping

Table 8 shows that usage growth in the developing world is rapid, but low density. Africa remains the big 'loser' in terms of internet use.

Table 8 Estimated internet use in 2007

World regions	% population (penetration)	Usage % of world	Usage growth 2000–07 (%)
Africa	4.7	3.4	882.7
Asia	13.7	38.7	346.6
Middle east	17.4	2.5	920.2
Latin America/Caribbean	22.2	9.6	598.5
Europe	43.4	26.4	231.2
Oceania/Australia	57.1	1.5	151.6
North America	71.1	18.0	120.2
World total	**20.0**	**100.0**	**265.6**

The benefits of the internet are numerous and it has a significant role in development, providing:

- education and health advice
- reduced isolation of rural areas

- access to the best price for buying and selling goods
- business opportunities such as online shopping
- communication with branch plants and customers
- 'technological leapfrogging' (the bypassing of technological stages that others had to go through)
- exchange of political ideas and information (although some countries like China, Myanmar and North Korea may restrict access to this information)

Switching on through tourism

The internet may have reduced the need to travel, but it seems to have increased the demand to travel. It allows 'exploration' of distant places, plus online booking of holidays, flights and tours. Tourism business has grown faster in the developing world than the developed, as lower cost air travel and a desire for 'something different' has led people to travel to more 'exotic' locations. Some developing countries gain up to 40% of their GDP from tourism. Many are tourist 'hotspots' that have seen rapid growth (e.g. Vietnam, Cuba and Madagascar).

Getting switched on through tourism requires less physical infrastructure than complex, hi-tech industries and uses natural, human and cultural resources already present in a region, but without careful management it can ruin the very resources it is based on.

Switching on and switching off: winners and losers

Developing regions often depend on commodity exports for their income. Bananas are a good example. There are two types:

- 'colonial' bananas — from small farms in former French, British and Dutch colonies in Africa and the Caribbean
- 'dollar' bananas — large plantation-grown bananas from Latin America sold by American TNCs (Dole, Del Monte and Chiquita)

In 1996 Banana Wars began because the EU favoured bananas from the small colonial producers and applied a 20% tariff to banana imports from dollar banana exporters. The UK and France argue that supporting small producers in the developing world is crucial to the economies of those small countries, while the USA, other dollar banana countries and TNCs that control 70% of the dollar banana trade disagree. The most recent WTO ruling, in April 2008, backed the dollar banana producers (the tenth WTO ruling on this issue).

This example shows that traditional connections based on links between nations are being replaced by connections created by TNCs. The Banana Wars have both winners and losers:

- Consumers in the EU are likely to get cheaper, better-quality, dollar bananas.
- The Big Three banana TNCs will extend their sales and profits.
- The dollar banana countries will see trade grow.
- The ex-colonial countries and small farmers are likely to see shrinking trade and income.
- For workers on the dollar plantations, jobs will continue, but pressure groups

argue that working conditions and wages are worse than in the colonial banana countries.

Switched off: the global periphery

Some peripheral locations have been bypassed by globalisation and remain switched off and seemingly trapped in poverty. Sub-Saharan Africa and parts of south Asia, central South America and the near east all fall into this category. The periphery has not attracted investment for a number of reasons:

- **Political instability:** wars, conflicts, poor governance and corruption deter investors due to high risks.
- **Debt** stifles economic growth and prevents investment in human resources through provision of education and health.
- **Lack of a good infrastructure** — efficient roads, rail, power and water supplies — means production costs are high and distribution is unreliable.
- **Physical constraints**, such as a harsh terrain and unreliable climate, combine with **environmental hazards** such as flooding and soil erosion.
- **Poverty and disease** produce a low-skilled, inefficient workforce that would cost too much to train.

The combination of these factors creates marginalisation. Where investment in the periphery does occur, through tourism and resource exploitation (mining, oil drilling), it is highly localised, creates few jobs and fails to produce connections that might generate broader economic growth.

Roots

Changes in UK population

Over the last decade, **net migration** (the difference between immigration and emigration) into the UK has been positive. This is unusual, as net emigration was the dominant trend for much of the twentieth century. The population is also ageing. The proportion of people over 60 is expected to rise to 26% by 2020 and 38% by 2050. The twin challenges of managing migration and a greying population are key issues for the UK.

Table 9 The UK's changing population

Statistic	UK figure (2006)	Trends	Explanation
Fertility rate (per woman)	1.84	Fertility has been declining, from 3.5 in 1900, to 2.95 in 1964, to an all time low of 1.63 in 2001. Since 2001 there has been a slight upward trend.	Women are having children later, many now focus on their career in their 20s. Later marriage is common, and children are seen as a cost. Availability of contraception and abortion.
Birth rate (per 1,000)	10.7	Birth rates have fallen from 28/1,000 in 1900. However, after the Second World War birth rates rose — the so-called 'baby boomer' generation.	Birth rates reflect fertility, and have fallen in line with it. Recent migration has helped the birth rate begin to rise; 20% of all UK births are now to migrants.

Statistic	UK figure (2006)	Trends	Explanation
Death rate (per 1,000)	10.1	The death rate trend has been smooth and steady, falling from 16/1,000 in 1900. Only the two world wars caused the trend to briefly reverse.	Improved healthcare, diet and education have driven the death rate down. Dangerous, dirty industrial jobs are a thing of the past.
Infant mortality (per 1,000)	5	There has been a drastic change in infant mortality, down from 140/1,000 in 1900.	This trend reflects the trend in death rates, with improved NHS postnatal care since 1945 having a significant effect.
Life expectancy (years)	81 for women 76 for men	Trends have been significantly upward, rising from 49 for women and 45 for men in 1900. The difference between men and women has existed for centuries.	Life expectancy has almost doubled as health has improved. Postwar vaccination programmes drove out diseases such as polio and TB and housing has dramatically improved.
Population age structure	Average age 39	Average age has risen, from 34 in 1971. By 2015, the number of people in the UK over 65 will exceed those under 16 for the first time.	Life expectancy improvements have driven up average age, but life expectancy is not likely to rise much further.
Family size	2.4	Family size has fallen, from around 6 in 1900 to 2.9 in 1971. 29% of UK households consist of one person, and 36% of two people. The most common family unit is a couple with two children, but this is becoming less common.	Many social changes have contributed to the decline in family size. Fertility decline is one, but also the rise in single people and people who never marry, and the decline in grandparents living within an extended family. The rise in divorce has created many more smaller households.
Ethnicity	8%	Some 8% of the UK population is non-white. This has grown from a tiny proportion in 1945. Around 50% are Asian/Asian British, with black/black British another 25%.	Postwar migration is the main driving force. This began in the 1950s and has continued since.
Total population	60.5 million	Total population in 1900 was 38 million. Population continued to grow steadily until 1971 (56 million) but since then growth has been slow.	The social and sexual revolution saw a marked shift toward lower fertility from the 1960s and this led to stagnation in the UK population in the 1970s. Recently population has begun to grow steadily again, partly as a result of increased immigration.

Researching your 'roots'

Either as a personal or class activity, you should research your own roots, as follows:

- Begin by profiling your own surname using: **www.nationaltrustnames.org.uk/**
- Conduct interviews with family members, focusing on where your family lived in the past and their employment history.

- Consider plotting changing family size over three or four generations.
- Compare your results with those of your peers to identify contrasts between urban and rural roots, and different ethnic or religious roots.

Employment and social status

One of the most significant changes in the UK has been the decline in primary and secondary employment, and the industrial transition. In 1920 there were 1.2 million coal miners in the UK; today there are just a few thousand. Even in the mid-1960s 1.5 million people were employed in the textile industry. Today 76% of the workforce is employed in the service industry, rising from 58% in 1975.

The transition to a service-based economy has been painful. In 1983 there were 3 million people unemployed (12% of the workforce) as the UK deindustrialised. By the twenty-first century the tertiary economy had created many new jobs, including highly paid professional jobs, and quaternary industry had grown. Only 10% of people owned their own home in 1900, compared with 68% today. Rising affluence has led to mass car ownership, foreign holidays and second homes. The sum of all of these changes has been to extend social mobility and increase the proportion of people in the UK who consider themselves 'middle class'.

UK population geography

Deindustrialisation and the move to the service industries has created considerable regional population change:

- internal migration to the south and east, averaging 30,000 people per year since the 1970s
- rural depopulation, particularly in more remote rural areas, as agriculture has mechanised
- a movement of people out of declining city centres, and into suburbs and the rural–urban fringe
- counter-urbanisation from cities to the accessible countryside, averaging about 90,000 people per year since the 1970s
- international immigration into cities, especially in the south, midlands and northwest

The greying population

The ageing, or greying, population has its roots in the 'baby boom' of 1945–65, when birth rates were higher than today. People born in 1945 will reach age 65 in 2010. Those born in 1965 will be 65 in 2030, when 25% of the UK's population will be over 65.

The ageing issue has been made more acute by 'Generation X', born in the 'baby bust' era (1965 onwards) as fertility rates fell dramatically due to female emancipation and the availability of the contraceptive pill. Generation X will be the working-age population in 2010–30 required to support the greying population (Figure 27).

In the future, **dependency ratios** (the ratio of over 65s and under 16s to the working-age population) will rise. In 1971 there were 3.6 working-age people to every pensioner, falling to 3.3 in 2003 and projected to fall to 2.3 by 2051. The ageing population presents us with a range of challenges and opportunities (Table 10).

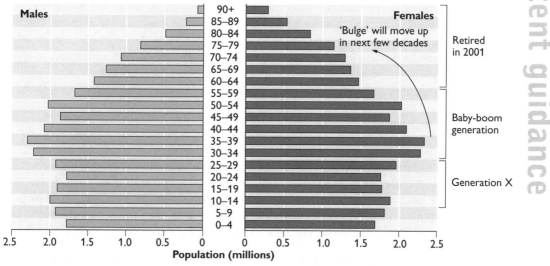

Figure 27 UK population pyramid, 2003

Table 10 The greying population: challenges and opportunities

	Challenges	Opportunities
Economic	Maximising employment (80%+) among the working-age population, to maximise tax revenue. Generally, and especially in ageing hotspots, the costs of providing long-term care will rise. NHS costs will increase, especially for expensive longer-term treatment of chronic conditions. General taxation, as a proportion of earnings, may need to rise. Skilled labour shortages may stifle economic growth and lead to inflation.	Many older people do not want to retire at 65; allowing them to continue to work would increase the tax take. Moving towards a system of personal pensions, for those who can afford it, would relieve pressure on the state system. Older people have experience, which employers should perhaps value more. The 'grey pound' may become a significant source of economic growth and new market opportunities. Many older people have both time and money, creating a huge market for age-specific products, from Saga holidays to domestic stairlifts.
Social	As people live longer, housing will remain occupied, creating demand for new housing for younger people. The number of single-person retired households will rise. Pressure to raise the state pension retirement age will grow, which may be resisted by younger people.	An older society may be a more law-abiding one, with less need for police and prisons. A desire to 'do good' and 'stay active' may lead to a rise in voluntary and community work. Older grandparents, still healthy and fit at 70 or 80, may take more responsibility for childcare.
Environ-mental	New sites will need to be found for care homes and retirement homes. Falling population could cause an increase in depopulation and dereliction in some areas.	Greying voters may become a powerful political force, possibly focusing on issues such as equality and environmental sustainability.

The UK may have to consider some radical actions in order to offset the ageing population:

- raising the retirement age
- using the tax system to encourage private pension provision
- using immigration to help raise fertility and avoid population decline
- changing attitudes so that older people are seen as assets rather than costs

On the move

In our globalised world, migration has increased and millions of people every year tear up their roots and move, legally or illegally, to new locations. This 'demographic globalisation' is a challenge for both source (origin) and host (destination) countries. In the UK, migration has increased the diversity of the population. If a country receives 500,000 immigrants in a year, but loses 300,000 emigrants, then the net immigration for that year is 200,000. Historically the UK is a country of net emigration (Figure 28).

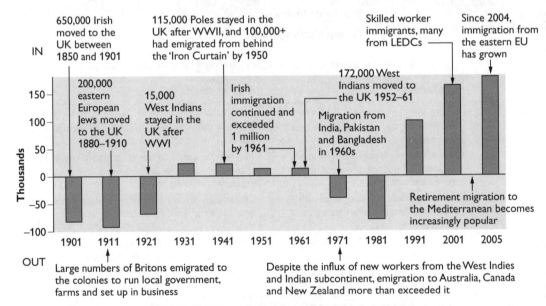

Figure 28 UK net migration balance since 1901

In the 1950s and 1960s the UK encouraged colonial immigration from the West Indies, India, Pakistan and Bangladesh (Commonwealth countries). This met an economic need for low-wage workers and explains the current concentration of these ethnic groups in northern industrial cities (Leeds, Manchester, Bradford) and London. Meeting its international obligations, the UK has also accepted refugees from war-torn parts of the world, including:

- 30,000 Asian Ugandans in 1972
- 20,000 Vietnamese boat-people in the 1970s
- 8,000 Bosnian refugees in the early 1990s
- 100,000 Somalis fleeing ongoing conflict since the early 1990s

Some refugees return home, but many cannot. Those who stay may apply for asylum and eventually become citizens.

Immigration has not been evenly spread across the UK. Some regions are much more diverse than others (for example, London is much more multiethnic than the north-east of England) and ethnic groups have concentrated in particular locations, such as Pakistanis in Yorkshire, Humberside and the west midlands, and black Africans in London.

Immigration has a role in total population numbers because:
- population numbers rise immediately if immigration exceeds emigration
- migrants tend to be young and fertile, so population is boosted by higher birth rates (in addition, some migrants may come from cultural backgrounds where larger families are valued)

In the twenty-first century, immigration has become a political 'hot potato'. There are several reasons for this:
- The number of people seeking asylum in the UK rose sharply from the late 1990s, peaking at 90,000 in 2002. Numbers had fallen by 75% by 2007 because of much tighter government policy.
- Immigration from the eight new eastern European EU member states since 2004 has resulted in the largest mass immigration in the UK's history. The number of eastern European migrants in 2008 is estimated to be around 1 million.
- There are growing concerns about the number of illegal migrants entering the EU (and ultimately the UK) across the 'porous' EU borders along the Mediterranean and in eastern Europe.

New migrant flows have created a sense of 'migration churning', which many people are uncomfortable with. This churning makes planning services such as schools, housing and hospitals challenging. The big question is whether, on balance, migration is a cost or a benefit.

Compulsory case study: eastern European migration

Eight new member states acceded to the EU on 1 January 2004: the Czech Republic, Estonia, Hungary, Latvia, Lithuania, Poland, Slovenia and Slovakia (the A8 or Accession Eight countries). The EU has an 'open border' policy for both trade and people. The 1995 Schengen Agreement removed border controls between many European countries. Migration to the UK by A8 countries has been on a much bigger scale than that to all other EU countries. In 2003 government estimates of the number of migrants who might move to the UK in 2004–05 were put at 20,000–40,000. In fact, over 200,000 immigrants arrived. This is explained by a number of **pull factors**:
- In 2004, A8 countries were not given the right of free movement except by the UK, Ireland and Sweden (the UK is not part of the Schengen Agreement and retains border controls), which narrowed the choices for potential migrants.
- The developed economy of the UK is a magnet for migrants seeking work and opportunities.
- The UK has a reputation for tolerance towards migrants.
- A growing, low-unemployment economy was in need of low-wage labour.

The intervening obstacles for new migrants were relatively minor because:
- a Worker Registration Scheme allowed migrants right of entry and the right to work
- cheap bus and air travel reduced costs
- existing established immigrant groups, such as the Poles, provided community and cultural support for newcomers
- immigrant-friendly services, such as 'Polish' UK bank accounts, were quickly set up, allowing remittances to be sent home

The number of migrants grew quickly from almost none in January 2004 to 665,000 resident migrants by late 2007. Most of the migrants work in low-paid temporary jobs in agriculture or services such as catering. They are spread across the UK, including rural areas such as the Scottish Highlands and Herefordshire. Table 11 outlines the key costs and benefits of A8 migration.

Table 11 Costs and benefits to the UK and A8 countries of post-accession migration

	Source countries (A8)	Host country (UK)
Costs	Loss of workforce: 7% in Poland and 10% in Lithuania. Brain drain and skills shortages: skilled migrants such as plumbers and doctors have left Poland. Ageing and decline: migration is reducing fertility at home; Poland's population could decline from 38 to 33 million by 2050. Social/age imbalance: as most migrants are young and male, this creates an imbalance at home which may affect traditional family units and relationships. Exploitation: some workers are exploited by gangmasters and are paid less than the minimum wage.	Social/cultural tensions, especially in rural areas with no history of immigration. Crime: a rise in low-level 'cultural unfamiliarity' crime, e.g. road traffic offences. Population growth: some projections suggest a UK population of 65 million by 2016; 20% of babies are born to recent migrants (8% of the general population). Downward pressure on wages for the low paid: there is some evidence that A8 migrants 'undercut' low-wage Britons. Pressure on space and housing causing locally rising housing costs; long term there may be pressure to build on greenfield sites. Locally there will be pressure on schools, the NHS and council services and additional costs linked to translational needs and the need for additional teacher support.
Benefits	Remittances: temporary workers send 25% of their earnings home. For Poland, this amounted to around €6.4 billion in 2006, or 2.5% of total Polish GDP. Higher wages: on average A8 migrants earn £6 per hour, much higher than at home, but lower than the UK average of £10. Skills can be taken back: returning migrants may have developed English skills, and other workbased skills and experience. Working A8 migrants abroad are better than unemployed people at home who require benefit payments.	Skills gaps filled: many industries such as fish processing in the Highlands and farming in the Fens have been 'saved' by low-cost labour. Economic turnaround: in some areas such as rural south Lincolnshire 10% of the population are A8 migrants meaning more local spending compared to pre-2004. Business opportunities: banks, supermarkets and other business have begun to provide 'Polish' food and other services. Counteracts ageing: if migrants stay, and have children, then the UK's population 'greying' may slow.

In 2008 the flow of migrants slowed. Approximately 1 million migrants have come to the UK, but around 350,000 have gone back home, illustrating that a large part of this migration is temporary economic migration.

The future
It is likely that eastern European migration to the UK will slow because:
- eastern European economies will develop over time, reducing the strength of the economic push and pull
- by 2007, A8 unemployment rates had halved since 2004, shrinking the pool of potential migrants
- by 2011, all EU countries must be fully open to A8 migration and many migrants will 'divert' to other countries

The A2 countries, Romania and Bulgaria, which joined the EU in 2007, had restrictions placed on free migration. These restrictions will be gradually lifted up to 2013.

Compulsory case study: Mediterranean flows
UK nationals increasingly emigrate south, towards sunshine and warmer temperatures. Around 750,000 Britons live permanently in Spain, and 200,000 in France. Total emigration reached 400,000 in 2007; approximately half of these emigrants were British (the rest were returning immigrants). These 'sunseekers' are very different from the A8 migrants. Most of them (60%) are economically inactive. Many are retired and some own businesses such as hotels.

Migrants to the Mediterranean are pulled there by a range of factors:
- housing, and the cost of living, is generally lower than in the UK
- as a growing tourist destination, the Med has many new business opportunities
- better weather is guaranteed — hot summers and mild winters
- lifestyle may be more relaxed, with a 'holiday atmosphere'

There are also push factors. Some are simply the reverse of the pull factors above (bad weather!) but many migrants cite factors in the UK such as:
- a perception of rising crime, declining respect and a sense that the UK has 'gone to the dogs'
- high taxation and spiralling house prices
- lack of space, congestion and low environmental quality

Retirement migrants may be able to sell a valuable house in the UK, buy a much cheaper one in Spain and have a substantial sum left over to add to their state/personal pension.

Other developments have made the process of emigration easier since the 1980s:
- Spain joined the EU in 1986.
- The Schengen Agreement and the 1992 Maastricht Treaty (which created the Eurozone) have made moving to another EU country much easier.
- Low-cost airlines provide easy and cheap access.
- Companies can sort out legal issues and paperwork employing English-speaking lawyers and estate agents.

- The internet allows easy communication with family members, friends and business interests back in the UK.

A criticism of 'sunseekers' is that they are not integrated with local communities and culture. This has intensified over time as English bars, doctors, schools, sports clubs etc. have set up, so there is even less need for emigrants to integrate with their Spanish or French hosts. There are extreme cases, such as San Fulgencio on the Costa Blanca: of its 10,000 population, 8,000 are foreign (half of these are British). The area was in the news in 2007 as British and German expats set up their own political party and stood in local elections, winning 21% of the vote and three local council seats.

This type of migration is on the rise. Concerns about mass emigration are being raised and emigrants tend to have mixed experiences, especially as they become infirm with age. Table 12 outlines the key costs and benefits of retirement migration to the Mediterranean.

Table 12 Costs and benefits of retirement migration to the Mediterranean

	Source country (UK)	Host country (e.g. Spain)
Costs	**Social** Family breakup, as grandparents move away. Loss of potential childcare. **Economic** Loss of a highly experienced workforce, especially if they retire early. The 'grey pound' is spent overseas.	**Social** Emigrant 'ghettos' are created, with little social and cultural integration. Resentment may grow as immigrants seek to enter local politics. House prices exceed the buying power of local people. **Economic** Some benefits and healthcare costs are borne by the host country. **Environmental** Large-scale villa development has ruined much of the coastal landscape and degraded biodiversity. Water supply systems are strained in semi-arid areas. Localised pollution has risen, and flood risk rises as urban development occurs.
Benefits	**Social** In part, emigration balances increased immigration, reducing net migration rates. **Economic** Fewer older people to take care of; some health and care problems are effectively exported. **Environmental** Relieves pressure for new homes, and therefore to build on greenfield sites.	**Economic** Increased spending in the local economy. Some retirees are highly affluent. Job creation in construction, retail and legal and health services. Areas which were largely unproductive scrubland become valuable land to build on.

Managing migration

Managing migration is a key issue for governments. Globalisation has pushed it to the top of most political agendas in the developed world. Migration policy is a tricky balancing act, as Table 13 shows.

Table 13 Managing migration: policies

Policy	Advantages	Disadvantages
Border controls	Physical borders, policing, passports and visas allow governments to count people in and out and so keep track of net migration.	These systems are extremely costly (the UK plans to introduce 'electronic' border control in 2010, at a cost of £2 billion). Tight regulations may put off some visitors and migrants who are needed to fill skills gaps.
Work permits	Work permit schemes allow temporary workers to be controlled and matched to skills shortages.	Work permits can be abused, with some not leaving when their 'time is up' and becoming illegal migrants.
Refugees and asylum seekers	Prestige is gained by accepting vulnerable groups and respecting basic human rights.	The public may perceive refugees and asylum seekers as a cost with few benefits attached. The asylum system is costly to administer.
Integration	Citizenships tests such as those in the UK and USA might help integration by expecting a basic understanding of language and cultural norms.	Critics argue that passing a test does not prevent social tensions and racism.

In the UK, migration policy has tightened recently as public concerns about mass migration have risen. UK government policy has moved towards an Australian-style points-based immigration system. The system needs to balance the economic need for skilled workers and younger people to counteract the greying population, while minimising the social and environmental costs of immigration.

World cities

A total of 3.3 billion people live in urban areas. In 2007–08 the number of urban dwellers exceeded the rural population for the first time.

- In 1900, 10% of people lived in cities; by 2030 this figure will rise to 60% of the global population and by 2050 it will be 75%.
- Half of the urban population today is under 25.
- Asia's urban population is set to rise from 1.4 billion to 2.6 billion by 2030.
- Africa's urban population will rise from 300 million to 750 million, and Latin America's from 400 million to 600 million.
- Some cities are considerably larger than major countries: Greater Tokyo (35 million) is larger in terms of population than Canada (33 million).

The definition problem

Deciding what constitutes a 'city' is far from straightforward. Even for a small provincial city such as Leicester there are problems. In 2006, 290,000 people lived within the city's administrative boundary, but some suburbs are in other administrative areas, so in fact the urban area had a total population of 440,000. One of the problems with population figures is that most of them are estimates, as censuses are carried out only every 10 years, and even they 'miss' some people (illegal workers, temporary migrants).

Urban agglomerations and megacities

It may be useful to consider urban areas and wider metropolitan areas, rather than cities. **Urban agglomerations** — similar to **conurbations** — exist when urban areas merge with each other. Urban areas with a population in excess of 10 million are defined as **megacities**. There were 20 megacities in the world in 2007, 15 of which were in the developing world. Many are expected to grow to reach 20 million people by 2015. New cities will also acquire megacity status. By 2020, or very soon after, there could be close to 30 megacities around the world.

There are 200+ cities with populations of over 1 million — around 70 of these are in China. These 'million cities' include many cities in the developed world that are stagnating or declining. A few lucky cities have attained 'world city' status. These cities have significant global economic and political power:

- **London** — global financial centre, world's busiest international airports, 23 Global 500 companies
- **New York** — global financial centre, 24 Global 500 companies, home of UN
- **Tokyo** — global financial centre, 52 Global 500 companies

Increasingly, megacities are transforming themselves into new urban forms. As cities grow, they reach a size limit (approximately 15 million people) beyond which the city cannot grow with one central business district (CBD). As this limit is reached the city becomes so congested that people begin to move outwards. Cities become more linear and a string of urban centres grows into a **megalopolis** or '**urban archipelago**'.

It is important to note that the rate of urbanisation is highly variable across the planet, with some cities growing at breakneck speed and others slowing to a crawl. There is a contrast in the nature of urban processes in large cities, depending on their location in the world and stage of economic development (Table 14).

Urban growth

The world's most rapidly urbanising cities have high population growth rates. Much of this is fuelled by rural–urban migration. This is most acute where rural poverty is highest. Food shortages, lack of land due to subdivision, fuel shortages, soil erosion, war and conflict, natural hazards — all push people out of rural areas. Perceived opportunities — jobs, better housing, healthcare, a chance of an education — are the pull factors.

content guidance

Table 14 World urbanisation types

	Urbanisation stage	Growth	Economy	Planning and cycle of urbanisation	Examples
Developing world	Immature	Very rapid: 3%+ per year. Largely migration growth.	Informal economy = 60%. Small-scale manufacturing, street trading, urban farming.	**Urbanisation** Little planning: uncontrolled sprawl. Squatter settlements dominate. Basic needs barely met. 60%+ live in slums. Environmental problems.	Kabul Lagos Kinshasa
	Consolidating	Rapid 2–3% per year. Balance of migration and internal growth.	Manufacturing important, some service industries. Informal economy around 50%	**Urbanisation and suburbanisation** Attempts at planning, focused on waste, congestion and water supply. Upgrading of slums and some social housing. Most basic needs met	Cairo Mumbai Jakarta Chongqing
	Maturing	Slow: under 2% per year. Largely internal growth.	Service industry dominates, with some manufacturing. Informal economy under 40%.	**Suburbanisation** Effective attempts at housing, transport and land-use planning. Environmental problems being tackled. Quality of life satisfactory for many. Gated communities in suburbs.	Mexico City São Paulo Beijing
Developed world	Established	Very slow, under 1% per year. Some are stable.	Dominated by professional, services and retail. Formal economy.	**Counter-urbanisation and reurbanisation** Large-scale suburbanisation, with counter-urbanisation. Since 1980, most have regenerated inner-city and former industrial areas. Quality of life is high for most, and environmental quality is good.	London San Francisco Paris Birmingham

Rural–urban migration has a double effect. Most migrants are young and fertile. This creates high natural increase, further fuelling population growth. In some developing cities, natural increase and rural–urban migration are contributing to staggering rates of urban growth. By 2015, every hour 42 people will be added to the population of Mumbai, 50 to Dhaka and 58 to Lagos.

In the developed world, trends are widely different:
- Greying populations have reduced natural increase, or even reversed it.
- Urban growth in cities like London is largely due to international immigration.

- Despite regeneration, suburbanisation and counter-urbanisation continue to draw people out of cities.

Because cities are at different stages of the cycle of urbanisation, life can vary considerably from one city to the next.

Rapid urban growth in the developing world has led to a slum crisis. Lacking jobs and forced to work in the informal economy, people build their own settlements. Most lack tenure (a requirement for improvements), and sanitation and clean water are patchy or absent. In Mumbai, for example, 60% of the population live in slums on 6% of the city's land. With a growth rate of 2.2% per year, the slum population will rise from 1 billion in 2005 to 1.4 billion by 2020.

Newcomers to developing world megacities inhabit 'old' slums such as Dharavi in Mumbai. These are tenement blocks and shanty towns with appalling overcrowding and sanitation. They have the advantage of being close to work in the city. Squatter settlements spring up on the 'septic fringe' through organised landgrabs. Slums are located where no one else wants to be — on dangerously steep slopes, next to polluted rivers, on marshland or downwind of polluting industry.

In both the developed and developing world the disadvantages of the inner city have created 'movers' who increasingly suburbanise to the urban edge. New suburbs are often 'gated' (i.e. walled off with security gates). This trend increases the segregation between rich and poor.

Managing cities

Megacities face a number of challenges depending on their stage in the cycle of urbanisation. In the developing world, problems usually result from a lack of resources; in the developed world, the problem is often overuse of resources and large urban ecological footprints. Both situations are unsustainable. At both ends of the development spectrum, a lack of management of urban waste and systems leads to environmental problems.

Environmental issues

In developed cities, ways need to be found to reduce inputs, for example through:
- water metering and pipe mending
- reduction in the use of packaging
- more public transport
- more efficient vehicles
- recycling building materials
- localised food distribution and improved storage

Outputs too must be reduced, by:
- using less polluting vehicles

- finding alternative energy sources
- recycling
- water reuse
- more carbon sequestration, e.g. urban gardens, farms and forests

Such methods must be implemented if overall ecofootprints are to decrease and the quality of life improve for urbanites.

Ecocities

China plans to build a series of ecocities to showcase urban sustainability. The first of these, Dongtan, is being constructed close to Beijing:

- It will be a largely car-free city.
- Energy will be provided from renewable sources — wind farms and solar panels.
- A waste treatment plant will convert sewerage and compost into biogas that will be used for cooking, heating and power generation.
- The city will comprise three compact districts, separated by parks, farms, lakes, pagodas and leisure facilities.
- Most people will live in apartment blocks six to eight storeys high, designed with natural ventilation to minimise the need for air conditioning.
- There will be two water-supply systems: one with drinking water and another providing 'grey water' for toilets and garden irrigation.
- Traffic lights will automatically switch to give priority to hydrogen-fuelled buses.
- The city will aim to be carbon neutral.

Improving slums

However, building from scratch in this way is not a viable solution for the millions of slum dwellers in rapidly expanding megacities. Help often comes from NGOs working in slum districts, with funding from local authorities, NGOs themselves and organisations such as the World Bank. Often, improving slums focuses on:

- providing residents with tenure (the right of ownership) as a first stage in giving them confidence to make improvements
- providing basic services, such as clinics and schools, often by city authorities
- NGOs providing improved water supply such as tubewells and standpipes
- residents themselves improving sewers, roads and houses through 'aided self-help' schemes, which provide building materials and basic training

There is a place for large-scale slum clearance and major sewerage and water engineering projects, but in many megacities the scale and continual growth of the slums defy this approach.

Global challenges for the future

The globalised world is increasingly:

- urban
- on the move
- connected

This is not the case for everyone, or even for the majority. Globalisation has created more choice but also more homogeneity — a Starbucks and a McDonald's in the centre of every city. Demographic globalisation has created mixing and diversity, but with it come the social and cultural tensions associated with rising migration. We are all global consumers but we live in a world fearing global climate change and worrying about resource shortages. Globalisation has a price. How can it be managed to reduce the costs?

A two-speed world?

Despite the gains from globalisation in countries such as China and India, the world continues to be widely unequal. In 1960 the ratio of the income of the poorest 20% of the world population to the richest 20% was 1:30. By the year 2000 this had risen to 1:70. In many parts of the developing world, the poor have been getting wealthier, but developed countries have been getting wealthier faster, thereby widening the gap. In many African countries, income levels have actually fallen since 1980.

Dramatic rises in consumption of resources are explained by global population growth (from 3.7 billion in 1970 to 6.4 billion in 2005) and higher consumption per person. Chinese consumption has risen dramatically and by 2031 may reach levels similar to those of the USA in 2005:

- 1,350 million tonnes grain per year (66% of world production in 2005)
- 180 million tonnes of meat per year (80% of world production in 2005)
- 2.8 billion tonnes of coal per year (global production 2.6 in 2005)
- 1.1 billion cars (800 million globally in 2005)

Goods that were once sourced locally, or were unavailable, are now imported by developing countries from overseas. This issue of 'food miles' — the distance that produce travels from farm to consumer — is important. The longer the distance, the bigger the need for refrigeration and for extra packaging, and the higher the 'carbon footprint' of that particular food. Some examples in the UK are:

- grapes from Chile — a distance of over 12,000 km, heavily packaged to avoid damage
- strawberries from Spain — a distance of approximately 1,600 km
- prawns from Indonesia — a distance of almost 12,000 km
- sending frozen Scottish-caught prawns to China for processing into scampi, then shipping these back to the UK for consumption
- sending coffee produced in Africa to India for roasting, then to the UK for sale

Exploitation or development?

The 'winners' from globalisation might be expected to be those that have gained jobs as TNCs have shifted production to low-cost locations — but is this true?

A typical factory worker in China's Pearl River delta has the following working conditions:

- Pay is $50–55 per month.
- Unions are banned.
- Overtime, a requirement to keep the job, is around 30 hours per week.

- Many migrant workers live in factory-owned dormitories, 16 people per room.
- Labour and health and safety laws are rarely enforced.

The alternative work is farming, which in China nets an annual income of $300–400 compared with $1,000+ for a factory worker. The majority of factory workers are rural–urban migrants who consider their lives improved compared with those of their parents. There are moves to increase worker rights and ban child labour in India and China. For governments there is a difficult balancing act as attempts to improve workers' conditions could simply mean TNCs shifting again, to the next low-cost location.

Green globalisation?

If a globalised world is to achieve sustainability, action is needed at a range of scales. There is a hierarchy of possible actions:

- individual actions — changing behaviour
- local actions — usually by local councils
- national — government policies
- global — agreements and targets

Concern over rising carbon emissions has moved up the political agenda and is now on the radar of many consumers. Those seeking to reduce their carbon footprint have a range of options (Figure 29).

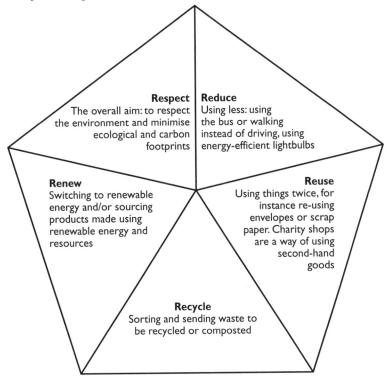

Figure 29 The 5 Rs

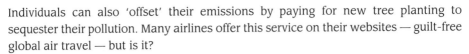

Individuals can also 'offset' their emissions by paying for new tree planting to sequester their pollution. Many airlines offer this service on their websites — guilt-free global air travel — but is it?

Governments can change consumer behaviour. Car tax (Vehicle Excise Duty, VED) is one of the UK government's ways of applying a 'green tax'. It aims to use the tax system to change our car-buying behaviour by linking VED to carbon dioxide emissions per kilometre. Introduced in 1998, variable VED does seem to have affected the types of car we buy.

Table 15 Taking action to create a more equitable world

Strategy	Example	Viability
Fair trade An attempt to reduce the economic unfairness of globalisation. By 2006, the total value of UK fair trade sales was £300 million and it is growing at 40% per annum.	Shoppers can choose to buy fair trade coffee and more of the money goes to the producers of the coffee beans. Other examples of fair trade produce are chocolate, bananas, wine and some clothing items such as jeans.	Buying fair trade goods means more money goes to poor workers. However, as the number of schemes grows, it becomes harder to monitor how 'fair' they are.
Ethically sourced goods Consumers can avoid purchasing goods produced under exploitative 'sweatshop' conditions. The work of the UK Ethical Trade Initiative (ETI) is worth researching.	Gap received bad publicity during the 1990s due to claims that Indonesian staff suffered poor working conditions. The firm has introduced stricter regulations for its overseas operations and guarantees workers 'dignity and respect'.	Outsourcing and supply chains among TNCs make codes of practice hard to enforce. Although firms such as Gap and Nike prohibit worker exploitation in their own factories, goods produced for them by a third party may have used 'sweatshop' labour.
Charitable donations and international aid Governments give aid and non-governmental organisations (NGOs) such as Oxfam and Christian Aid collect money directly from the public to help address the economic unfairness of globalisation.	In 1984–85 and again in 2004–05 Band Aid raised money for famine relief in Ethiopia. Around £100 million was raised in the 1980s. In 2004–05, money was also given to people living in Sudan's Darfur region.	Aid can result in dependency for poorer nations and can make it difficult for emerging businesses to profit. In Zambia, clothing manufacturers have gone bankrupt due to the free second-hand clothes donated by OECD charities — known locally as salaula.
Trade reforms Governments and international lobbying organisations have tried to improve terms of trade for poor nations, especially rules regulating the import and export of agricultural produce. Protesters gather at World Trade Organization and G8 conferences and make their case for change.	Huge subsidies paid to European farmers under the Common Agricultural Policy (CAP) and protective trade tariffs encircling the European Union (EU) force up the cost of imported African goods. Changing these rules would help African farmers.	The Commission for Africa has drawn attention to the need for reforms of subsidies, tariffs and non-tariff barriers for poorer countries. European farmers resist measures that open markets up to greater competition, as this could threaten their livelihoods.

Many businesses, even TNCs, have jumped on the 'green' bandwagon as public concern has grown. In 2002 BP changed its company logo from a shield to a flower, and the company has diversified into solar power and retailing. Many people are cynical about the motives of business for 'going green' but there are reasons to suggest companies such as BP could be serious:

- Money can be saved by reducing energy use (and therefore pollution).
- Company assets are vulnerable to sea-level rise and flooding linked to climate change.
- Oil and gas will run out; renewable energy provides an alternative business model.
- Consumers are increasingly 'switched off' by companies with a poor environmental record.

If companies can see long-term benefits to shareholders of 'going green', they are likely to move in this direction.

Ethical globalisation?

Ethical concerns over the nature of globalisation have led some consumers to change their behaviour in the hope of reducing the negative social consequences of consumption (Table 15).

Questions
&
Answers

In this section of the guide, there are four questions — two Section A questions and two Section B questions.

The Section A questions are on global hazard trends and global challenges for the future. Each question is accompanied by outline answers and comments on how marks will be awarded, along with some exam tips on how to improve your performance in the exam. A ✔ is used to indicate points for which a mark would be awarded.

The Section B questions are on the impacts of global warming and globalisation. Each is accompanied by suggestions for what to include in a good answer and outline mark schemes. A sample C-grade answer is also included for each question, along with examiner comments, preceded by the 🅔 icon, on how to improve it.

Section A

Global hazard trends

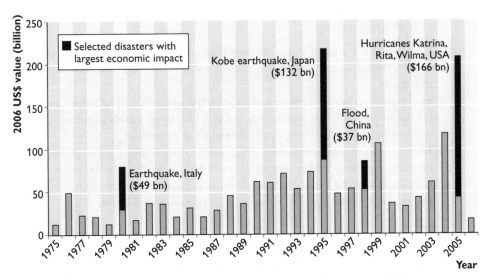

Figure 1 Annual reported economic damages from natural disasters, 1975–2006

(a) (i) Define the term 'natural disaster'. (2 marks)

   (ii) Which is the correct statement concerning trends in natural disasters?

     A There have been increasing numbers of tectonic disasters in recent years.

     B The number of deaths from disasters has decreased dramatically in recent years.

     C The relative amounts of damage are greater in more developed countries than in developing countries.

     D The number of people affected by natural disasters has increased tremendously in recent years. (1 mark)

(b) (i) Describe the trends shown in Figure 1. (2 marks)

   (ii) Give two examples of economic damages. (2 marks)

   (iii) Explain with an example why certain natural disasters prove to be extremely costly in terms of economic damages. (4 marks)

Total: 11 marks

question

Answers and commentary

(a) (i) To gain both marks, you need to explain both 'natural' *and* 'disaster' (realisation of a hazard) (1 mark for each term).

(ii) The correct answer is D. For A, only hydro-meteorological disasters are increasing, tectonic ones are **fluctuating**. For B, the decrease in the number of deaths has levelled off, not decreased. For C, relative economic damage is greater in developing countries.

(b) (i) Generally upward trend ✓ but with significant anomalies in certain years ✓, showing fluctuations ✓. Each of these points is worth 1 mark, up to a maximum of 2 marks.

(ii) Examples of economic damages include infrastructure loss, loss of houses and possessions, loss of agricultural crops and loss of tourism installations. Each of these points is worth 1 mark, up to a maximum of 2 marks.

(iii) Here you should name a disaster ✓ that is likely to be a high-magnitude disaster (more damage) ✓ in a more developed country ✓ (value of insured goods) and in a highly populated area ✓. An example could be Hurricane Andrew or Katrina. Each of these points is worth 1 mark, up to a maximum of 4 marks.

Exam tips

- Pay attention to the mark weightings.
- Always try to use examples to support your statements.
- Always use data from the resource.
- Always use correct terminology in your answer.

Global challenges for the future

Study Figure 2, which shows ethical behaviour in the UK in 1999 and 2007, and answer the following questions.

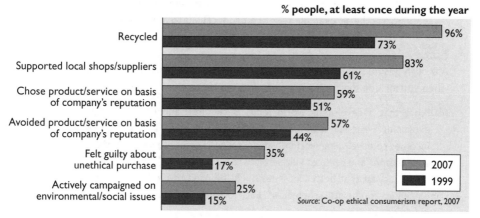

Figure 2 Ethical behaviour in the UK, 1999 and 2007

(a) Which behaviour involves trying to reduce the environmental impact of household waste? (1 mark)

(b) Identify two changes in behaviour between 1999 and 2007. (2 marks)

(c) Suggest why 35% of consumers in 2007 'felt guilty about an unethical purchase'. (3 marks)

(d) Explain how fair trade might benefit producers in the developing world. (4 marks)

Total: 10 marks

■ ■ ■

Answers and commentary

(a) Recycling (1 mark)

(b) All behaviours became more common/saw an increase ✔. Some behaviours changed by around +20% ✔, e.g. recycling, supporting local shops and feeling guilty. Others increased only slightly, by around 10%, e.g. campaigning and choosing/avoiding products ✔. Each of these points is worth 1 mark, up to a maximum of 2 marks.

Exam tip

Make sure you quote data from the graph and look for patterns, in this case the fact that some behaviours have risen sharply while others only gradually.

question

(c) Possible suggestions include:
- Certain products may have been made using developing world labour.
- Certain products may be bad for the environment.
- There is growing awareness generally of movements such as fair trade.

You will receive 1 mark for a correct general statement, plus a further mark if it is supported by an example (e.g. of a product), up to a maximum of 3 marks.

Exam tip

Although this question does not ask for examples, as a general rule you will gain credit when you include them.

(d) Note that a developing world focus is required and there is only 1 mark for explaining what fair trade is. At the most basic level fair trade provides higher incomes to developing world producers ✔; it creates greater certainty about prices and incomes and reduces vulnerability to fluctuations ✔. Better answers should go on to suggest that better incomes might be used to fund community projects ✔, e.g. education/health schemes ✔. Each of these points is worth 1 mark, up to a maximum of 4 marks.

Exam tip

Most candidates would be able to say that fair trade raised producer incomes, but this would only gain 1 mark. In the exam, when you have written an explanation, ask yourself 'so what?' In this case, the 'so what?' is that the higher incomes could lead to improved diets and health ('so what?'), so people could work harder, raising incomes further and so on.

Section B

The impacts of global warming

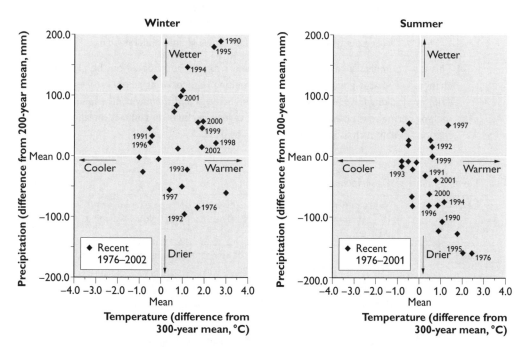

Figure 3 *Variation in precipitation and temperature compared to the long-term means, 1976–2001, UK*

(a) **Study Figure 3. Analyse the patterns shown in the winter and summer variations in precipitation and temperature in the UK.** (10 marks)

(b) **With reference to the Arctic examine the environmental impacts of global warming.** (15 marks)

Answers and commentary

(a) Winters are in general getting wetter (26 are above the mean) with only nine drier than the mean. They are also getting warmer (only seven not warmer) but there are high variations.

Summers are less variable but there is a definite trend towards drier (all except 11) and also warmer (all accept 11) summers. These changes are almost certainly linked to global warming.

Mark scheme

Level	Mark	Descriptor
Level 3	8–10	An accurate analysis, with good use of detail. Shows good understanding of resources, well structured, recognises link to global warming.
Level 2	5–7	Some analysis of the data showing satisfactory understanding of the graph. Begins to describe pattern with some structure.
Level 1	1–4	Attempts to analyse patterns, but shows limited understanding.

(b) This is a compulsory case study so examiners expect good detail on the precise changes — what is going on in terms of melting sea ice, rising temperatures, etc. They are looking for specific impacts on the ecology and the environment such as melting permafrost, poleward migration of vegetation belts, and impact on land and marine animals such as ice seals and polar bears.

Mark scheme

Level	Mark	Descriptor
Level 4	13–15	Wide-ranging, well-structured account. A detailed examination of a range of environmental impacts. Well related to Arctic global warming. Well written.
Level 3	9–12	A sound account with a range of environmental impacts described. Structured response showing good links to Arctic environment.
Level 2	5–8	A general account, with some descriptions of several impacts. Geographical terminology infrequently used, sometimes in error. Some structure.
Level 1	1–4	Basic description of one or two impacts, poorly structured, with frequent language errors.

Answer to part (a): C-grade response

(a) In winter, the graph shows that <u>many</u> ✓ of the temperatures are warmer than the 300 year ✓ mean, and that in <u>quite a few years</u> it seems to be wetter especially in <u>1990</u> which was the warmest and wettest year. ✓

In summer the pattern is not very clear to see for the temperature with some years a little warmer and some <u>a little colder</u>. However it does seem to have got much drier in recent years ✓, with a lot of warmer and drier summers. ✓

The changes shown are very much the result of climate change ✓ which has led to rising temperatures. ✓

> 🄴 This answer would just reach Level 2. Ticks indicate valid points and underlining shows weaknesses. The graph has been understood and correctly interpreted. However the answer is far too brief and it does not really make use of or analyse the data. Numbers of years are not given but vague terms like 'many' are used, and the date of only one extreme is given. It would have been useful to mention how

much some individual years deviated from the mean. For example 1976 had summers that were 2.5°C warmer than the 300 year mean, and over 150 mm drier.

The comment on climate change could be expanded to consider the impacts of global warming and the reasons for linkage to wetter winters.

It would also have been helpful to comment on the general distribution of data in more detail.

Answer to part (b): C-grade response

There are numerous environmental impacts of global warming in the Arctic.

Increasing temperatures have led to the coniferous forest belt moving ✓ north to where the tundra used to be, the result is that tundra ecosystems are being lost.✓ The increase of forests increases the risk of forest fires. ✓ Also the warmer temperatures allow the spread of diseases such as the spruce bark beetle. ✓ The changes in the Arctic have had a huge impact on animals such as seals ✓ and polar bears who because of the melting ice have lost their habitat. Also the rising temperatures have led to the melting of the permafrost ✓ which has released a lot of methane into the atmosphere. As many native peoples rely on the Arctic ecosystems for their traditional way of life the lack of species ✓ to hunt is affecting their way of life. The rising temperatures are also having an impact on the oceans because rivers are bringing a lot of extra fresh water into the Arctic Ocean. ✓

🄴 This answer just reaches Level 3. The following improvements would have gained it an A grade.

First, it is vital to put the Arctic in context and to say exactly what the changes in the Arctic are. You must state the scale and rate of the changes resulting from rising temperatures, for example giving temperature rise in °C and facts and figures on the rate of melting of ice sheets and sea ice. You should know this from the case study.

Second, the above answer is short on terminology such as food web, thermohaline circulation, positive feedback loops, albedo.

Third, the impacts given in the answer are largely negative. Positive impacts such as the migration northwards of cod fishing, the benefits to farming of a longer growing season, or the improved sea access for trade and tourism (due to lack of sea ice) should be mentioned.

Fourth, it is a good idea to include some place-specific detail, for example of the names of rivers going into the Arctic or the names of native peoples, such as Inuit.

Globalisation

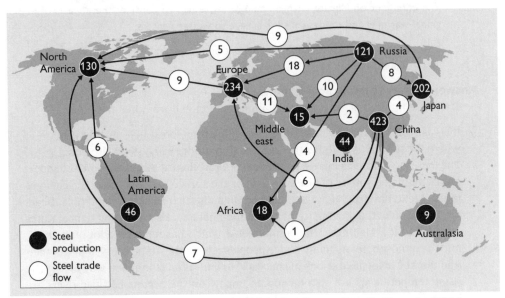

**Figure 4 The global network of steel production and trade flows in 2006
(data in millions of tonnes)**

Study Figure 4, which shows the global network of steel production and
trade flows in 2006, and answer the following questions.

(a) Explain the uneven pattern of global steel production and trade flows. (10 marks)

(b) Using examples, examine the role of technology in the spread of
globalisation. (15 marks)

Answers and commentary

(a) Good answers might include the following key points:
- Trade is a fraction of production in all regions. This suggests transporting steel
  may be very expensive.
- There is some evidence of export to regions that are geographically close, e.g.
  Europe and Russia to middle east. Again, this could reflect high transport costs.
- Some regions which lack industrialisation are very dependent on imports, i.e. the
  middle east.
- Production is heavily concentrated in Asia, with China (423 million tonnes) as
  the largest producer, around double any other. China is a low cost NIC producer
  (global shift).
- China is a net exporter to higher cost locations, e.g. Europe and the USA, where
  production costs are higher.

- North America is a large producer, but also a large net importer as production costs are lower elsewhere. This might suggest that production is still close to the TNC home country.
- Steel production and imports are lowest in the least developed regions, i.e. Africa, as there is little demand for it.
- You could also discuss the presence or absence of iron ore in the different locations.

Mark scheme

Level	Mark	Descriptor
Level 3	8–10	Detailed explanations for some flows and some patterns; shows good understanding of global network concept. Uses examples and appropriate geographical terminology to show understanding. Minimal written language errors.
Level 2	5–7	Some explanations for some patterns and flows. Becomes descriptive at the lower end but with some geographical terminology used. Some written language errors. Becomes descriptive at the lower end but with some geographical terminology used. Some written language errors.
Level 1	1–4	Little structure. Describes some aspects of the map without any real reference to geographical context or use of terminology. Frequent written language errors.

Answer to part (a): C-grade response

(a) The pattern is that production is concentrated in certain areas of the world. The biggest producers are North America, Europe, Japan and China. They produce about 200–300 million tones each year. These are rich countries which control global steel production. However, places like China are industrialising and developing and need steel. Some TNCs might have moved steel production to places like China because production costs are lower than in, say, the UK for instance. Africa is a very poor country and has very little need for steel due to most people living in slums so it has very small steel production and flows. The flows are complicated with some places exporting and others importing. Russia is a big exporter and so is Latin America. North America is a big importer. As said previously, Africa is not really involved in production and trade. Some of the pattern could be due to trade blocks, for instance, there could be agreements allowing countries in Asia to import and export with no barriers, quotas and tariffs.

e The data need to be used more precisely and more often to support the explanations of specific patterns. Errors such as the classic 'country of Africa', must be avoided. Terms such as global shift need to be included to show precise understanding. Trade flows are covered only generally and there is limited explanation and vague description. Mention of a specific trade bloc (not block) such as ASEAN would add detail and realism.

■ ■ ■

(b) The focus of your answer should be on the role of technology, although some drift away from the role of technology into other factors would be acceptable as the question does say 'examine'. Good answers might include the following key points and are likely to come to a brief conclusion:

- The internet is an obvious example. It allows companies to communicate easily and cheaply with branch plants, customers and suppliers and plays a key role in allowing offshoring. Individuals can 'explore' holiday destinations online. Mobile phones are similar in terms of 'keeping in touch' with widely dispersed networks.
- Air travel has shrunk the world (e.g. 747, Concorde, A380) since the 1950s, with budget airlines bringing the 'shrinkage' to the masses; this has allowed cultural mixing and migration as never before; it also aids the globalisation of TNCs.
- Containerisation is responsible for a revolution in transport costs and is a key factor in allowing TNCs to shift production globally. Bar code and other tracking software and devices allow complex companies like airlines and parcel carriers to operate globally.

Mark scheme

Level	Mark	Descriptor
Level 4	13–15	Structured, detailed account which focuses on the role of technology. Uses appropriate terms and range of examples to show understanding. Likely to be evaluative in style. Written language errors are rare.
Level 3	9–12	Structured and uses some examples to explain the role of technology with appropriate geographical terminology to show understanding. Minimal written language errors.
Level 2	5–8	Some structure. Descriptive account which has some explanation of the role of some technology; likely to have narrow focus, but with some geographical terminology used. Some written language errors evident.
Level 1	1–4	Little structure. Describes one or two technologies with no real reference to globalisation context or use of terminology. Frequent written language errors.

Answer to part (b): C-grade response

(b) Technology has played a very important role in globalisation and will continue to do so. The internet was invented in the 1980s and was beginning to be used commonly by the mid-1990s. Today up to 1 billion people use the internet globally. The internet allows companies to communicate with employees and suppliers all over the world and this encourages companies to locate abroad (called outsourcing), knowing they can still easily contact people. The internet has allowed many TNCs to locate in the free trade zones in China and keep in touch with them. The internet is costly though, so not everyone in the world has it. In poorer developing countries they may not have satellite and telephone networks, internet service providers, computers etc. Satellites are very important as these allow the internet to work, but most are controlled by TNCs or developed world governments which might explain why the internet has not spread to many very poor countries.

This means the internet is not available and this will discourage globalisation in these places. Africa needs the internet if it is to develop and this is why the '$100 laptop' has been introduced. The internet is very important in globalisation. TNCs are the next most important technology as some are more powerful than countries. General Motors is as rich in turnover as Bulgaria. This gives TNCs the power to move production around the world and they play an important role in spreading globalisation.

e This answer is unbalanced. It focuses too heavily on the internet; reference to a range of technology is needed to earn a higher grade. TNCs are not a technology; the section on TNCs is not irrelevant, but it needs to be worded more carefully. Some appropriate terminology is used, such as outsourcing, but it is not used with enough precision. Some examples are included, such as the $100 laptop, but more depth and detail is needed.